Fundamentals of Water Finance

Fundamentals of
Water Finance

Michael Curley

CRC Press is an imprint of the
Taylor & Francis Group, an **informa** business

CRC Press
Taylor & Francis Group
6000 Broken Sound Parkway NW, Suite 300
Boca Raton, FL 33487-2742

Printed on acid-free paper
Version Date: 20160819

International Standard Book Number-13: 978-1-4987-3417-2 (Hardback)

Library of Congress Cataloging-in-Publication Data

Names: Curley, Michael, author.
Title: Fundamentals of water finance / Michael Curley.
Description: Boca Raton, FL : CRC Press, 2017. | Includes bibliographical references and index.
Identifiers: LCCN 2016025967 | ISBN 9781498734172 (hardover : alk. paper)
Subjects: LCSH: Waterworks--Finance. | Municipal water supply--Finance. | Water-supply--Finance. | Sewage disposal--Finance. | Water treatment plants--Finance. | Sewage disposal plants--Finance. | Water quality management--Finance.
Classification: LCC HD4456 .C873 2017 | DDC 363.6/10681--dc23
LC record available at https://lccn.loc.gov/2016025967

Visit the Taylor & Francis Web site at
http://www.taylorandfrancis.com

and the CRC Press Web site at
http://www.crcpress.com

Printed and bound in the United States of America by Publishers Graphics, LLC on sustainably sourced paper.

Contents

SECTION III The New Game in Town

SECTION IV Resources and Disclosure

Acknowledgments

In 1993, I wrote a prequel (if you will) to this book called *The Handbook of Project Finance for Water and Wastewater Systems*. The Acknowledgments section of that book starts with these words:

> First and foremost, I must thank George Ames of the United States Environmental Protection Agency (EPA). While on a partial sabbatical from EPA, during which he founded the Council of Infrastructure Financing Agencies (CIFA), George asked me to make several presentations on financing for small water and wastewater systems. He then asked if I would commit my presentation to writing. This became *Financial Alternatives for Small Water and Wastewater Utility Systems*, which was published as a monograph by CIFA in January 1990, and which is the source document for this book. If it were not for George, this book would never have been written.

The same is true of this book that you are now reading. George and I have remained close friends for the past 28 years. And again, "If it were not for George, this book would never have been written."

The next people to thank are Scott Fulton and Jay Pendergrass, who are the president and vice president for programs and publications, respectively, of the Environmental Law Institute in Washington, DC, where I am a visiting scholar. They have always been encouraging of my efforts at research and writing about environmental finance.

Next, I need to thank my partner, Brent Fewell, in our new "virtual" law firm called the "Earth and Water Group." Brent gave me lots of good advice and a lot of help with the editing.

Next is my dear friend of 25 years, Kristin Franceschi, who is one of the most remarkable people I know. Kristin is a senior partner in a major US law firm, DLA Piper. She is not only a bond lawyer but also a former president of the National Association of Bond Lawyers. Kristin is a world-class ballroom dancer and goes to dance competitions monthly. Finally, Kristin is a recovering Stage IV cancer victim—colon cancer! In Chapter 21, you will read about the role of bond counsel in utility finance transactions. You will see that I use the pronouns "she" and "her" in this section. This is in Kristin's honor.

There are three other key people at the Environmental Protection Agency (EPA) whom I need to thank: first, Peter Grevatt, the head of the Office of Ground Water and Drinking Water; Kelly Tucker, who is the greatest source of information on non-point source pollution (he is with the Clean Water State Revolving Fund (CWSRF) Branch at the EPA); and finally, Mark Mylin, who is the keeper of all the numbers and the history of the CWSRF.

I also need to mention Joe Mysak, my friend of 35+ years. Long ago, just after the earth cooled, I founded the third municipal bond insurance company in the world on Wall Street. Joe was then the publisher of *The Bond Buyer*. We had lots of fun together, and I learned a lot from Joe. Since then, Joe has gone on to be the editor of the daily *Bloomberg Brief*, "Municipal Market." And along the way, he wrote the *Encyclopedia of Municipal Bonds* and *Handbook for Muni-Bond Issuers*. I am still learning from Joe.

Here are some more of the many special people I need to thank for their help with this book:

- Jag Khuman of the Maryland Water Quality Financing Administration and Jim George of the Maryland Department of the Environment
- Paul Marchetti of PennVest
- Jim Maras of USDA's Rural Development
- Tracy Mehan and Deidre Mueller of the American Water Works Association
- Sam Wade and Matt Holmes of the National Rural Water Association, Ruth Hubbard of the Minnesota Rural Water Association, Elmer Ronnebaum of the Kansas Rural Water Association, and Gary Larimore of the Kentucky Rural Water Association
- Michael Deane and Marybeth Leongini of the National Association of Water Companies
- Adam Krantz and Amber Kim of the National Association of Clean Water Agencies
- Eileen O'Neill and Lori Harrison of the Water Environment Federation
- Diane VanDe Hei and Carolyn Peterson at the Association of Metropolitan Water Agencies
- Julia McCusker of CoBank

Finally, I need to thank my friend Colin Bishopp of Renew Financial, who tirelessly travels the country trying to teach the world how to use the Clean Water Act and the CWSRF to finance energy efficiency and renewable energy projects in homes and businesses across America.

And to the many people whom I am sure I have forgotten, thank you too.

Finally, I need to mention my mother, Lucy A. Curley, who for 20 years was the treasurer of the city of Buffalo, New York. She taught me the principles and the importance of finance.

Introduction

Why should the financing of water (and wastewater) projects be any different—or more challenging—in the twenty-first century than they were in the last century? Let me answer this question with one of my favorite stories.

When I was growing up in Buffalo, New York, on the shores of Lake Erie, I could walk into the water up to my neck and look down and see my toes. A few years later, when I was in law school, I would walk into the water up to my neck and look down. I could barely see my shoulders. Now, when I go back, I can walk into the water up to my neck and see my toes again.

When I was a boy, municipal sewage was fouling the Great Lakes. Cities and towns all along the shore were dumping untreated or poorly treated sewer water into the lakes. These devils are largely gone. Now we face a much more insidious gang of pollution villains.

Congress finally caught on to the sewage treatment problem and—after much hand-wringing and debate—finally passed the Clean Water Act of 1972 (CWA), which President Richard Nixon promptly vetoed. Much to their credit, the Senate overrode Nixon's veto the same day. The House overrode it the next day, October 18, 1972, when it became law.

What saved the Great Lakes was the construction grant program embedded in the Clean Water Act. This provided massive grants to sewage treatment utilities all across the country. So now, Chicago, Milwaukee, Detroit, Cleveland, Buffalo, Rochester, and many other cities received billions of dollars to finance upgrades of their sewage treatment plants. By the time the construction grant program was repealed in 1987, during President Ronald Reagan's administration, it had given out over $70 billion in grants. State and local governments had matched much of this money to the tune of probably another $30 billion, so about $100 billion in 15 years.

Ronald Reagan was not a fan of government grants, but he could live with loans. So Congress replaced the construction grant program with the Clean Water State Revolving Loan Fund (CWSRF). Despite its awkward name, this program is actually a clean water bank. Congress has appropriated money for this program each year since 1987. EPA distributes it to the states. States are required to match it on a 1:5 basis. They then use it to make water pollution control loans.

The cool aspect of this program—and the reason Congress called it a revolving loan fund—is that as the individual loans are repaid, the money revolves back into the state CWSRF and stays in this fund. It does not go back into state treasuries. Furthermore, as a condition of accepting the federal capitalization grant each year, the states must promise to keep the SRFs financially intact and whole. They cannot raid the SRFs for money when the state tax coffers get low. This means that when SRF loans are repaid, these funds are available to make more water pollution control loans. In addition, all of the states, except Vermont, charge interest on their loans. None of them charge market rates. They all offer subsidized interest rates. But charging low interest rates is better than charging no interest rates. The point is that as these CWSRF loans are repaid, both the principal and the interest go back into the fund. Since 1987, the federal government has contributed over $40 billion to the CWSRF. States have contributed another $10 billion. But because of the revolving nature of the program, the CWSRF has been able to provide over $111 billion of financial assistance since 1987. So for every $1 Congress contributed, they received about $2.80 of projects. Reagan would be proud.

Fine. So what is different about the twenty-first century? Remember the Lake Erie story? The $100 billion from the construction grant program and the $111 billion from the CWSRF have done their job. That is why I can see my toes again. One hundred percent of the construction grant program funds went to sewage treatment projects. And over 96% of the CWSRF funds have also gone for sewage treatment facilities. They won the battle. Sewage is no longer the number-one source of water pollution in the United States.

Now, the number-one polluters in the United States are nonpoint sources.

And that is why the twenty-first century will be much different for those in the business of water pollution control. Nonpoint source pollution projects are many orders of magnitude more difficult to finance.

This book is divided into four sections. Section I deals with the basic mechanics of traditional project finance. There are nine chapters. There is a lot of good information there.

Section II deals with traditional water and wastewater finance programs—from the massive US tax-exempt municipal bond market to some small finance programs for equipment replacement and the like, all run by state rural water associations. All of these finance programs are for treatment works, collectors, mains, pipes, and so on, the usual twentieth-century stuff.

Section III of this book is titled "The New Game in Town." This is about the new initiatives to pay for a host of new types of projects that deal with a host of new nonpoint source pollution problems. Some of the broad categories you will see are your soon-to-be new friends such as stormwater projects, green infrastructure, agricultural runoff projects, groundwater protection, watershed preservation, energy efficiency (yes, energy efficiency in a water finance program!), and the latest entry—resiliency projects. You will also see some new initiatives here, such as public private partnerships (P3s), "pre-project funding," small community special assistance, and "sponsorship" programs. Many of these involve exciting—but unfamiliar—new concepts.

The fourth and final section of the book is entitled "Resources and Disclosure." This section deals with whom you can turn to for advice and assistance on financing both new and traditional projects as well as where you can go to get training and networking with peers to help you develop your approaches to dealing with project finance, especially financing these new nonpoint source projects. Finally, disclosure is not a major factor in financing water and wastewater projects. There is much to learn about this relatively new subject.

Such matters as disclosure, credit enhancement, and subsidies bring new perspectives to many of the traditional water and wastewater finance programs. And, as you will see, financing some of the new projects is a fascinating new world of its own.

Author

Michael Curley is a lawyer who has spent most of his career in project finance and the past 25 years in energy and environmental finance. He joined the Environmental Law Institute as a visiting scholar in 2013. He is the author of *Finance Policy for Renewable Energy and a Sustainable Environment* (Taylor & Francis Group 2014). He also wrote *The Handbook of Project Finance for Water and Wastewater Systems* (*Handbook*), which was published by Times/Mirror. He contributes to the *Huffington Post* and is a member of the American Society of Journalists and Authors and the National Press Club, where his sits on the Book and Author Committee. He has published over 40 articles.

In 1990, Curley was appointed to the Environmental Financial Advisory Board at the US Environmental Protection Agency, where he served for 21 years under four presidents. Over the past 20 years, he has taught environmental and energy finance and law courses at Johns Hopkins University and George Washington University as well as the Vermont Law School. He founded the Environmental Finance Centers at the University of Maryland, Syracuse University, and Cleveland State University.

In the early 1980s, he raised venture capital for, founded, and served as president and CEO of the third municipal bond insurance company in the world and the first to insure economic development projects. He sold the company to a major international bank. He was also a partner in the New York City law firm of Shea & Gould.

He has served in several roles in government: first, as deputy commissioner and general counsel of the New York State Department of Economic Development and then as president and CEO of the state's bank for economic development. He was also the general counsel of the New York State Science and Technology Foundation, the state's venture capital agency. Before that, Curley served as parliamentarian of the New York State Assembly and associate counsel to the Speaker and was also an assistant to Congressman Richard D. McCarthy (D-NY) in Washington, DC. He also served as an adjunct professor of banking and finance at New York University, teaching venture capital and capital markets and was appointed to the board of directors of the United Nations Development Corporation by New York City Mayor Ed Koch.

For the US Environmental Protection Agency, Curley served as senior financial advisor to the Office of International Affairs, where he developed national municipal bond banks for water, wastewater, district heating, and other municipal infrastructure for both the Russian Federation and Ukraine, designed a revolving fund for water resources for the Ministry of the Environment in the Republic of Georgia, and built 15 rural water systems and developed the national model for rural water finance in the Republic of Kazakhstan.

As a member of the board of the International Rural Water Association, Curley built water and basic sanitation systems in Honduras, El Salvador, and Guatemala. He continues this work privately now in Guatemala. In late 2016, he and a colleague at the Bloomberg School of Public Health will start writing a book about bringing safe water and basic sanitation to the rural poor in developing countries.

Section I

Traditional Finance Programs

1 The 800-Pound Gorilla!!!

When it comes to financing water and wastewater (w/w) projects in the United States, the municipal bond market is the 800-pound gorilla! The size of this market is $3.7 *trillion*. Yes, *trillion*!

About $30–$40 billion of w/w projects are financed each year. About 90%–95% of these funds come—one way or another—from the municipal bond market.

Of this 90%–95%, about 85% comes from the market "directly". This means that w/w systems issue municipal bonds, themselves, to fund their projects. The other 5%–10% comes from municipal bonds that are issued by 50 state agencies (plus Puerto Rico) that manage the Safe Drinking Water State Revolving Fund (SDWSRF) and/or the Clean Water State Revolving Fund (CWSRF), collectively known as the SRFs. Let me explain.

As you will see in Chapter 3, the SRFs are funded with a grant by Congress each year. To be eligible for this grant, each state must match the federal grant on a 1 : 5 basis. These federal SRF grants are called "capitalization grants." That's because they capitalize the SRFs' "funds." This money cannot be handed out as grants; it can only[*] be loaned. It must "revolve." Here's how it works:

Your system applies for a loan from your state's SRF agency. The money "revolves" out to you. As you pay it back, it "revolves" back to the state agency. Then, another w/w system applies for a loan and the money then "revolves" out to that system, and so on. You realize that banks have been doing this "revolving loan" thing since the fourteenth century. However, instead of calling these institutions "Clean Water Banks," Congress decided to call them "revolving funds." Go figure.

So, the SRFs each have the money that they received from Congress plus the matching funds that they received from their own state. In addition, the SRFs also have interest income. Before loans are made, the SRFs keep their funds in interest-bearing accounts at commercial banks. The interest earned from these accounts goes into the SRFs' general funds. Plus, when they make you a loan, they charge you interest. Not much interest, mind you; all 50 SRFs make subsidized loans.[†] However, the interest you pay them back also goes into the SRFs' accounts.

Finally, in states that face strong demand for their SRF loans, the state agencies also issue municipal bonds to get more money to lend. These loans are backed—not by their respective states—but by the money in the SRFs' general accounts, both present and future, that is, the projected future payments on outstanding loans are also pledged. Using the present and future money in their general funds to raise even more money from the bond market is called "leveraging." Thirty-one states have leveraged SRFs.

So, between going directly to the municipal bond market and going indirectly to that market by going through a leveraged SRF program, that's where 90%–95% of the money comes from to fund your projects.

---

[*] With very limited exceptions, as explained in Chapter 3.

[†] Vermont, alone, charges 0%.

So, where does this money come from? From where does the municipal bond market get money?

First and foremost is the general public. People—just regular folks like your aunt and your grandmother—invest in w/w projects by buying municipal bonds that are issued by w/w systems. They do so because it is a good deal. Water and wastewater bonds are one of the safest investments anyone can make. They almost never default. In addition, as you will see in Chapter 2, most have extremely high credit ratings. For all that, they generally pay a pretty good return. So, for people who are interested in a decent return on a really safe investment, you can't beat municipal bonds for w/w projects. In addition, the interest you earn on most municipal bonds is exempt from federal income taxes, as well as some state income taxes.*

The second source of money for the municipal bond market is mutual funds and other institutions that want tax-exempt income. Property/casualty insurers, for example, invest the premiums you pay them. They pay income taxes on the interest they earn on those investments. If they want to avoid paying income taxes, they can invest in tax-exempt municipal bonds. Depending on the net difference between a higher-interest-paying, taxable investment and a slightly-lower-interest-paying tax-exempt investment, they make their decisions.

Mutual funds—tax-exempt mutual funds—just save your aunt and grandmother the trouble of deciding on investing in individual bonds. Should they invest in an AA-rated California school district bond or an AAA-rated bond issued by a local water authority in Oklahoma? If they invest in a mutual fund, they let the fund managers make those decisions.

In addition to the municipal bond market and the 51 SRFs, there are a few more federal government programs, one quasi-federal government program, some state programs, and some programs run by state rural water associations.

The US Department of Agriculture's Rural Development has a grant and loan program for communities with less than 10,000 population. This amounts to about $1 billion a year.

CoBank, the National Bank for Cooperatives, makes loans to w/w systems that are owned by cooperatives.

The Economic Development Administration and the Department of Housing and Urban Development occasionally makes grants or loans to public w/w systems.

Finally, a small number of projects are financed otherwise, mostly through local bank loans or through finance programs run by state rural water associations.

So, as you can see, if you want to know how to finance projects for w/w systems, you need to be familiar with municipal bonds, since they are—by far and away—the largest source of funds for such projects. They are, indeed, the 800-pound gorilla of the w/w project finance business.

So, this book begins with what municipal bonds are, how they are issued, and what you need to know about this process.

---

* Privately owned w/w systems can also issue bonds; but they are not tax-exempt.

2 Municipal Bonds

In Chapter 1, we said that municipal bonds financed the vast majority of water and wastewater (w/w) projects. This chapter will discuss municipal bonds. But, first, let's start with: What is a bond?

There are many kinds of bonds in the dictionary. A bond is something that holds things together, like bonds of affection or matrimony, or bonds of steel. In insurance, you have surety bonds, which is money that will be paid if someone doesn't do what he's supposed to do. In the criminal justice system, there are bail bonds, which, like a surety bond, refer to money that will be paid if someone doesn't do something he's supposed to, which, in this case, refers to showing up in court.

For our purposes, however, a bond is an "instrument of indebtedness" or "evidence of a debt." In simple terms, a bond is a form of loan. In fact, a bond is nothing more than a loan that is (1) on a standard form document and (2) where the borrower has a published—or otherwise well-known—credit rating.

Think of a $1000 US Savings Bond. All the language on all $1000 bonds is the same. It is, of course, a standard form. In addition, the US government has a AAA rating from two of the three international credit rating agencies.

Furthermore, when you buy a $10,000 bond from the state of Texas, which is part of, say, a $100,000,000 bond issue, you would find that each such $10,000 bond had exactly the same wording. Each bond is the same. In addition, as you could easily learn, Texas has a triple-A rating from all three credit rating agencies.

This concept of standard language has broader implications. A w/w bond from a system in Florida won't have exactly the same language as a w/w bond from a system in Oregon, but it would be close enough. Both organizations are in the same business. Both derive revenues from the customers they serve, and so on. So, w/w bonds from all over the country look pretty much the same. All school district bonds look pretty much the same, and so do municipal library bonds, parking garage bonds, and so on.

In addition to standardized documentation, bonds have published credit ratings, which are, usually both the rating on that individual bond issue (i.e., "bond issue," not individual bond)* and the rating of the entity issuing the bond. You can look up the credit rating on any bond you want to buy or can just ask your broker. There are three major rating agencies that publish credit ratings of major corporations, governments, and government agencies. In almost all cases, the debtor, or the borrower, will have a published credit rating. So, a bond is just a loan on a standard form to a borrower with a known credit rating.

Therefore, a "municipal" bond is just a standard form loan to a nonsovereign—that is, state or local government or government agency—whose credit rating is published and well-known. When a government borrows like this, it is called "issuing a bond."

---

* The exception is where a highly rated issuer issues a "special revenue bond," where the special revenue stream is not seasoned or strong.

So, municipal bonds are loans to state and local governments or agencies used to build public projects or acquire public property. Most municipal infrastructure in the United States is financed with municipal bonds. As noted in Chapter 1, the municipal bond market in the United States is more than $3.7 *trillion*. Even in these depressed times of local government borrowing, a little less than $350 billion of municipal bonds were issued in 2014 and a little more than $400 billion bonds were issued in 2015. Because of the cost and complexity of government programs such as the Clean Water State Revolving Fund, most major water utilities with good credit ratings go directly to the municipal bond market. In 2014, more than $30 billion of w/w projects were financed through the municipal bond market. That's about normal. The market has never had a problem absorbing this supply, which is less than 9% of its annual volume. In addition, the 2 SRFs supply about another $5 billion a year, much of which comes from the municipal bond market.

The first municipal bond in the United States was issued by the City of New York in 1812. In 1913, when Congress adopted the income tax, they exempted the income on municipal bonds from state and local obligations. This feature makes the American municipal bond market unique. In many countries, local governments, or local authorities, issue bonds. Only in the United States, the income from these bonds is exempt from federal income taxation.

Now, nothing is simple in life, especially when government is involved. So, it should not surprise you to learn that not all municipal bonds are tax-exempt and that, of the ones that are tax-exempt, some are not necessarily exempt from all forms of federal taxation.

There is a type of municipal bond called a "private activity bond," the income on which is NOT exempt from federal income taxation.

According to the Municipal Securities Rulemaking Board, a "private activity bond" is

> a municipal security, of which the proceeds are used by one or more private entities. A municipal security is considered a private activity bond if it meets two sets of conditions set out in Section 141 of the Internal Revenue Code. A municipal security is a private activity bond if, with certain exceptions, more than 10% of the proceeds of the issue are used for any private business use (the "private business use test") and the payment of the principal of or interest on more than 10% of the proceeds of such issue is secured by or payable from property used for a private business use (the "private security or payment test"). A municipal security also is a private activity bond if, with certain exceptions, the amount of proceeds of the issue used to make loans to nongovernmental borrowers exceeds the lesser of 5% of the proceeds or $5 million (the "private loan financing test"). Interest on private activity bonds is not excluded from gross income for federal income tax purposes, unless the bonds fall within certain defined categories ("qualified bonds" or "qualified private activity bonds"), as described below. Most categories of qualified private activity bonds are subject to the alternative minimum tax.
>
> "Exempt facility bonds"—These are private activity bonds issued to finance various types of facilities owned or used by private entities, including airports, docks, and certain other transportation-related facilities; water, sewer, and certain other local utility facilities; solid and hazardous waste disposal facilities; certain residential rental projects (including multi-family housing revenue bonds); and certain other types of facilities. Enterprise zone and recovery zone facility bonds are also considered exempt facility bonds.

As you might imagine, municipal bonds issued by privately owned w/w systems are "private activity bonds." They are, however, "qualified," so the interest on them is tax-exempt. But there's a catch. Qualified private activity bonds are subject to volume caps that are assigned under the Internal Revenue Code to each state. The states dole out allocations under the volume caps annually. Most w/w projects take 2–3 years for planning and design. So, the private water companies can never be sure whether they'll get an allotment when it's time to go to market with their bonds. For this reason, the private water companies, and their trade association, the National Association of Water Companies, and the American Water Works Association have all been urging Congress to exempt these types of bonds from the volume cap requirements.

In addition, to make life even more complicated, although the income from "qualified private activity bonds" is exempt from federal income tax, it is subject to the alternative minimum tax. If you really want to know more about this subject, ask your bond counsel or your municipal financial advisor.

There are two broad categories of bonds: General obligation (GO) bonds and revenue bonds. These are important for w/w systems.

"Units of local government"—Cities, counties, towns, townships, and villages—have very general powers to raise revenue. These entities can issue bonds that are backed by the local government's "full faith and credit." This is an unconditional promise to repay the debt from any and/or all of the sources of revenue that the government is authorized to raise. These "full faith and credit" bonds are the GOs, or GO bonds.

Units of local government can also issue revenue bonds. In such case, the local government promises to repay the bonds "only from specific revenues" that are spelled out in the bond documents. These are revenue bonds. It is common for governments to pledge that they will raise sufficient revenues each year to repay the bonds. So, if the w/w service provider is a local government, then its bonds will be as described above.

Many w/w service providers are not general units of local government but rather are local government agencies or public authorities. These agencies and authorities issue two types of municipal bonds: General revenue bonds and special revenue bonds.

General revenue bonds are issued by agencies that have multiple sources of income. Think of a sewer authority that charges its normal, annual rates for its usual services. Then, it also has 5-year surcharges for connection fees. It also sells services to septic tank maintenance companies and buys and sells property and lots of equipment. If such an agency issues a "general revenue bond," it means that it is pledging all of its revenues from all of these different sources to the repayment of its bond. Note that this is very much like a GO bond issued by a local government, where the local government pledges all of its revenues from all of its various sources to the repayment of the bonds.

A "special revenue bond" is just what you might think. In the case of our sewer agency, it might be a bond where it is financing extensions on its system and where it only pledges the specific income it receives from connection fees.

As you can tell, municipal bonds appear to be quite complicated. However, in practice, they are not complicated. If they were mind-blowingly complicated, there wouldn't be $30–$40 billion of them issued each year. So, if you get yourself a good bond lawyer and a good financial advisor, you should have no trouble.

3 The "Value" of Money

THE TIME/VALUE THEORY OF MONEY

Everyone realizes that things cost less years ago. They even begrudgingly admit that things will cost more in the future. However, how it all works is a bit of a mystery. Measuring these changes in value seems somehow like black magic.

If you know for certain how much an object cost a specific number of years ago and you know exactly how much it costs today, then you can calculate the rate of change going backward. The rate of change going backward is called the "discount rate."

You can also make predictions based on discount rates. For example, you can say something like this: "If a widget cost X 20 years ago, and costs Y today, and if the rate of change remains the same, then it will cost Z in another 20 years."

When you are talking about a "going foreward" rate, it is generally called the "compounding or compound rate."

All of this discounting and compounding is part of "The Time/Value Theory of Money." It is not a theory at all; it is a hard fact. Costs "do" change over time.

Most costs rise at, or close to, the rate of inflation. Some costs, for example, college tuition and medical care, seem to rise at rates greater than the rate of inflation.

The rate of inflation is compiled by the US Bureau of Labor Statistics from observing actual prices of a large number of goods and services. The bureau reports that the average rate of inflation in the United States from 1914 until 2013 was 3.35%. So, let us call it 3% for the sake of simplicity.*

We are now going to present a series of three examples to illustrate how the value of money changes over time.

The first, and, perhaps, most familiar example is the savings account.

The second example deals with a US savings bond and illustrates how the surrender value is calculated.

The third example uses a 30-year fixed-rate mortgage to show the subtle changes, which are actually occurring, when neither the home nor the payment appears to change at all over the 30-year period.

Next, we will present some actual illustrations of how the change in the value of money over time relates to the payments a water or wastewater system makes on behalf of its projects.

Now, let us consider the first example: the savings account.

Let us say that you have $10,000 that you want to put into a safe investment account to have for your retirement. Let us say that you find a fixed-income mutual fund that will offer you a guaranteed rate of 5%, compounded annually. How much money will you have in 10 years?

---

* Corporations often use their own discount number that they estimate as their profit goal. They sometimes call this their "opportunity cost" or "hurdle rate."

Before we go any further, please note that the terms "present value" and "future value" are relative to each other. They are relative to each other. "Present value" refers to the value that is earlier in time. "Future value" refers to the value that is later in time. So, when we calculate the price of a quart of milk 50 years ago, today's price is the "future value" and the price 50 years ago is the "present value."

Here is how we calculate the value of your retirement account in 10 years:

Year	Value ($)
1	10,000 × 1.05
2	10,500 × 1.05
3	11,025 × 1.05
4	11,576 × 1.05
5	12,155 × 1.05
6	12,763 × 1.05
7	13,401 × 1.05
8	14,071 × 1.05
9	14,774 × 1.05
10	15,513 × 1.05
	16,289

There you have it. Your $10,000 retirement account will have grown to $16,289 in 10 years at a "compound" rate of 5% per year. There is a formula for this exercise. It is

$$FV = PV * (1+r)^n$$

This is the compounding formula. Here "FV" is, of course, the future value and "PV" the present value. The r is the interest rate and the n is the term of years.

The opposite of the compounding formula is the discounting formula, which is

$$PV = \frac{FV}{(1+r)^n}$$

This formula can also be expressed as follows:

$$PV = FV * \left(\frac{1}{(1+r)^n} \right)$$

In other words, getting from the original $10,000 to the future value of $16,289, you multiply the amount by 1 plus the interest rate, each year. Getting back from the future value of $16,289 to the present value of $10,000, you simply take the $16,289 and divide it by "1 + r" or 1.05, each year.

Now, let us take an example of a US savings bond.

This is what happens when someone gives a baby a $1,000 US savings bond to help pay for its college education. Savings bonds mature in 20 years. So, we know that we can redeem the bond for $1,000 in 20 years. What is it worth today?

Well, using the second discounting formula, we would first add the rate of interest (.05) to the number 1 to arrive at 1.05. Then, we would divide the number 1 by 1.05 to get 0.9524. Then, we would multiply $1,000 by 0.9524, twenty times. The answer, which is the value of the bond today, is $376.89. So, the price of the baby's bond today is $376.89.

Our third and last example involves a 30-year home mortgage. There are two lessons in this last example.

Let us say that you took out your 30-year, fixed-rate mortgage in 1990 for $150,000 at a rate of 6% for a home you purchased for $190,000. As such, your first monthly payment in 1990 was $899.33; call it $900. Let us also say that you were earning $50,000 a year back then. This means that your yearly house payments (12 × $900 = $10,800) equaled about 20% of your gross salary of $50,000.

Now, your last payment in 2019 will also be $900; but by then, you should be making about $121,000 a year.* Your salary went up, but your house payment remained the same. So, in 2019, your yearly house payments will only comprise about 9% of your gross salary. This is lesson number one. Here, the value of your work (i.e., as it is reflected in your salary) changed. It grew over time, and as it grew, the cost of your home, which was fixed, became a relatively smaller and smaller part of your earnings. So, as your earnings grew, the burden of your house payments shrank. This is illustrative of how the value of money changes over time, and you can measure it against any sum that is fixed over time, which, in this case, is your mortgage payment.

The second lesson involves the value of your home. Let us say that in 2019, the kids will be grown and gone, and it will be time to sell the house and downsize. So, you will sell your home in 2019.

If the value of your home inflated at the same pace as the growth of your salary, then in 2019, it should be worth about $461,000.

So, now you should be familiar with the concept that the value of money changes over time. So, you should not be surprised to learn that "in 1990 dollars,"† your last mortgage payment is not worth $900 at all. In fact, it is worth only $149.44. This is just like the example of the savings bond. There, we knew what the future value was ($1,000), so we could calculate back to the present value ($376.89). Here, too, we know the future value of the last payment. It is $900. So, we can calculate back to determine its value 30 years before, which is $149.44.

So, as you can see, the time/value theory of money has serious implications for the w/w industry. It says that future payments will be less (i.e., be a smaller fraction of a growing income stream) over time. Moreover, as you will see in Chapter 8, it argues for financing projects over the longest period of time possible.

---

* Compounding and discounting throughout the chapter is at a rate of 3%.

† Phrases such as "1990 dollars" are just the short-hand ways of talking about a relative value at a given time.

4 Types of Long-Term Debt

TYPES OF DEBT

Types of debt are classified by their principal repayments. There are four types of loans or four methods of repaying the principal of a loan:

- Level payment method
- Level principal method
- Balloon payment method
- Irregular payment method

We are not going to spend a lot of time on either balloon payments or irregular payments. They are rarely used by water and wastewater (w/w) utilities.

Irregular payment bonds/loans occur sometimes when there are a mix of repayment sources—with different terms—for the debt. For example, a utility might issue a bond to pay for an expansion of its treatment plant *and* for, say, 200 new connections. In such case, the utility might charge the homeowners connection fees to homeowners, over a 5-year period. On the other hand, the plant expansion would be paid for by all ratepayers over a 30-year period. So, the bond payments would be substantially different in the first 5 years than that in years 6–30 of the bond term.

Balloon payments involve bonds that have normal repayment schedules until the last payment. For example, a balloon payment bond of $1 million might have a 5-year term with a 20-year level principal payment amortization schedule. This means that in each of the 20 years, the annual principal payments would be $50,000 each. However, at the end of the 5-year term, the final payment of $750,000 is due and payable. The $750,000 is the "balloon."

THE LEVEL PAYMENT METHOD

In the United States, most municipal utility debt is level payment debt. This is very suitable for utilities that have fixed rates (tariffs) that they charge their customers, since the payments are the same each year.

The level payment method means that "the total amount of principal plus interest paid is the same every year." As the amount of interest declines with every payment that is made, the amount of the principal paid increases by the same amount.

Calculating annual debt service payments (ADSPs) for level payment loans involves the use of a complicated formula, as follows:

$$\text{ADSP} = P * \left(\frac{i}{(1-(1/(1+i)^n))} \right)$$

Handheld business calculators can deal with this formula very easily with four keystrokes. However, for those without such devices, the ADSP on a 40-year loan can take a long time to calculate. Just to illustrate how the formula works, here is the calculation of a 2-year loan of $100 at 5% interest. There are seven steps involved. Here is the legend:

ADSP: Annual debt service payment
P: Original principal amount of the loan
i: Interest rate expressed as decimal
n: Term of the loan

Here are the seven calculations:

1. $\text{ADSP} = P \times (i / (1 - (1 / (1 + i)^n)))$
2. $\text{ADSP} = \$100 \times (0.05 / (1 - (1 / (1 + 0.05)2)))$
3. $\text{ADSP} = \$100 \times (0.05 / (1 - (1 / (1.05)2)))$
4. $\text{ADSP} = \$100 \times (0.05 / (1 - (1 / 1.1025))$
5. $\text{ADSP} = \$100 \times (0.05 / (1 - 0.90702))$
6. $\text{ADSP} = \$100 \times (0.05 / 0.09298)$
7. $\text{ADSP} = \$100 \times 0.53775$
   $\text{ADSP} = \$53.76$

You can easily prove that this is correct. You know that 5% interest on $100 is $5. So, of the first year's payment of $53.76, you know that $5 of this amount must be interest and the rest, $48.76, must be principal. If you pay off $48.76 of principal in the first year, there is only $51.24 of principal outstanding on the loan at the beginning of the second year. The second year's payment must be $51.24 of principal plus 5% interest. Five percent of $51.24 is $2.52. So, principal of $51.24 and interest of $2.52 equal the ADSP of $53.76. So, now you know that the formula is correct.

Going forward, for the purposes of illustrating the various principal payment methods, let us assume a 5-year loan of $100,000 with an interest rate of 10%.

To find the ADSP, we simply plug the variables into the formula.

$$\text{APSP} = \$100{,}000 * \left(\frac{0.10}{(1-(1/(1.10)^5))} \right) = \$26{,}380$$

Once we have the ADSP, we can now create an ADSP schedule to determine exactly how much interest and principal are paid each year. An ADSP schedule is also called an "amortization schedule."

Here are the rules for creating an amortization schedule for the level payment loan in our example:

Creating an Annual Debt Service Payment Schedule for the Level Payment Method

1. Determine the annual payment (ADSP) using the equation above.
2. Calculate the first year's interest payment (Interest = $i \times$ principal).

3. Obtain first year's annual principal payment by subtracting the interest from the ADSP of year one (Principal = ADSP – interest).
4. Subtract the year one principal payment from the original principal amount.
5. Calculate the second year's interest payment (Interest = $i \times$ outstanding principal balance).
6. Repeat the above process for each year of the loan term.

Here is what this process looks like in a tabular form.

Year	Prior Balance ($)	Interest ($)	–	Total Annual Payment (ADSP) ($)	=	Principal Payment ($)
1	100,000	10,000	–	26,380	=	16,380
2	83,620	8,362	–	26,380	=	18,018
3	65,603	6,560	–	26,380	=	19,819
4	45,783	4,578	–	26,380	=	21,801
5	23,982	2,398	–	26,380	=	23,982
Total		**31,899**	–	**131,899**	=	**100,000**

Summary of the Level Payment Method

- The annual payments are always the same.
- The amount of principal paid each year increases by exactly the same amount by which the interest payment decreases.
- The total of all the annual principal payments equals the original principal amount of the loan.

THE LEVEL PRINCIPAL PAYMENT METHOD

The level principal payment method means that the total amount of principal is the same from year to year and the interest payment is simply calculated using the remaining balance. As the amount of interest decreases with every payment that is made, the total ADSP also decreases.

Calculating ADSPs for loans using the level principal payment method:

P: Original principal amount of the loan
n: Term of the loan

Notice that the annual principal payment is the original principal amount divided by the term (P/n).

In the case of the example of a 5-year loan of $100,000 at 10% interest, the annual principal payment would look like the following:

$$\text{Annual principal payment} = \frac{\$100,000}{5} = \$20,000$$

The ADSP would be different in each year. However, in the first year, it would be the interest payment, which is 10% of the original principal balance of $100,000, or $10,000 plus the annual principal payment of $20,000. So, the first year's ADSP would be $30,000.

Here is what the full amortization schedule would look like in a tabular form.

Year	Prior Balance ($)	Interest ($)	+	Principal Payment ($)	=	Total Annual Payment (ADSP) ($)
1	100,000	10,000	+	20,000	=	30,000
2	80,000	8,000	+	20,000	=	28,000
3	60,000	6,000	+	20,000	=	26,000
4	40,000	4,000	+	20,000	=	24,000
5	20,000	2,000	+	20,000	=	22,000
Total		**30,000**	+	**100,000**	=	**130,000**

Here are the rules for creating an amortization schedule for level principal payment debt.

Creating an Annual Debt Service Payment Schedule for the Level Principal Payment Method

1. Calculate the level annual principal payment by dividing the original principal amount by the number of years in the term of the loan (Annual principal payment = P/n).
2. Calculate the first year's interest payment by multiplying the original principal amount of the loan by the rate of interest (Interest = $i * P$).
3. Obtain the first year's annual payment by adding the level principal payment to the first year's annual interest payment (1st year's ADSP = interest +principal payment).
4. Obtain the outstanding principal balance at the end of the first year by subtracting the first year's level principal payment from the original principal amount of the loan (Outstanding balance = P – ADSP).
5. Calculate the next year's annual interest payment by multiplying the outstanding principal balance from the preceding year by the rate of interest (2nd year's interest = outstanding balance * i).
6. Calculate the next year's annual payment by adding that year's annual interest payment to the level principal payment (2nd year's ADSP = 2nd year's interest + principal payment).
7. Continue the process for each year of the term.

Summary of the Level Principal Payment Method

- The annual principal payments are the same each year.
- Both the annual interest payment and the annual payment decrease each year.

- The total of all the annual principal payments equals to the original principal amount of the loan.
- Note that if there were no interest (an interest rate = 0%), then the level principal method would be the same as the level payment method, that is, the payments would be the same each year.

Comparison of the Level Payment Method and the Level Principal Payment Method

There are three very good reasons why the level payment method is preferred by US utilities. First, as you can see, with the level principal payment method, the amount of the payment declines each year. This makes little sense. Payments come out of the ratepayers' pockets. If the ADSP on a project is X this year and X – Y next year, there will be excess money floating around next year. True, there might be some justification in building up a so-called "rainy day" reserve; but, in general, ratepayers won't be too happy watching their utility amass large amounts of cash over the years that is coming out of their pockets.

The second reason relates to the "value of money," which we discussed in Chapter 3.

With the level principal payment method, the largest payment is due today—when money is more expensive than it ever will be over the coming years. So, why pay off so much with today's expensive money?

The third reason is planning. With the level payment method, the payment is the same every year. It facilitates your utility's financial planning.

So, in these first four chapters, you have learned that most w/w projects are financed by tax-exempt municipal bonds and that the tenor of almost all these bonds is that they are repaid according to the level payment method.

5 Comparing Financing Alternatives

This chapter will deal with comparing loans or bonds.

First, we will demonstrate how to compare simple loans. Then, we will look at how to compare loans when other fees or charges are involved. These are like the "points" that the banks charge on home mortgages as well as attorney's fees, appraisal fees, servicing fees, and the like. However, first, we will start with comparing the two most common forms of loans: level payment loans and level principal payment loans.

COMPARISON OF LOANS

There are two ways of comparing loans or bonds: the annual payment method and the total payment method.

The Annual Payment Method

In terms of comparing loans, the annual payment method works only with level payment loans. However, it is very powerful.

The annual payment method consists of calculating the annual payments on the level payment loans you are working with and then comparing them. It is very simple when using the formula in Chapter 4, or better, using a calculator that knows the formula.

The annual debt service payment (ADSP) on a level payment loan (Loan A) of $10,000,000 at 6% for 10 years is $1,358,680. The ADSP on a level payment loan (Loan B) of $10,000,000 at 4% for 5 years is $2,246,271.

Loan A = $1,358,680
Loan B = $2,224,271

Certainly, there is no problem in comparing these two loans by the annual payment method. The next question is, if you were a member of your utility's board, which method would you choose? What if your system had to have a major rate increase in the last year or so and now you are faced with this—an additional rate increase? Would you choose the $1.3 million rate increase or the $2.2 million rate increase? Easy question, right? An important part of your job on the board is to hold down the costs your customers have to pay. Does it concern you that Loan A is 5 years longer than Loan B? Probably not. Does it concern you that the interest rate on Loan B is 50% lower than that on Loan A? Again, probably not.

As you can see, the annual payment method of comparing loans is very useful, very powerful, and, fortunately, very simple.

The Total Payment Method

The total payment method looks easy, but it is not. You would think that all you would have to do is add up the annual payments on two loans and then look at them. So, in the example above, we would just multiply the $1,358,680 level annual payment in Loan A by its term of 10 years, or $13,586,800. Then, we would multiply the $2,224,271 level annual payment in Loan B by its term of 5 years, or $11,121,355. You would then compare them, right?

Loan A = $13,586,800
Loan B = $11,121,355

Comparing them this way is easy, isn't it? Yes. However, it is wrong. This is not a comparison of the total payments of these two loans. We have just compared apples with oranges.

Unfortunately, in order to correctly compare any two loans, you must discount each of their annual payments. Discounting was discussed in Chapter 3.

Let us see how we can compare financing alternatives by discounting their annual payments.

Let us say that the water system, which you are on the board of, has a $1,000,000 project. Let us say that you have three friendly local bankers who would like to lend you the money.

Banker A will lend money to your system for 5 years at a rate of 8% with level principal payments.
Banker B will lend money to your system for 5 years at a rate of 8% but with level payments.
Banker C will lend money to your system for 10 years at a rate of 9% and also with level payments.

Let's look at these loans. Here are the annual payments on Loan A.

Year 1	$280,000
Year 2	$264,000
Year 3	$248,000
Year 4	$232,000
Year 5	$216,000

To obtain the annual payment on Loan B, we must use the formula we saw in Chapter 4 because it is a level payment loan:

$$\text{ADSP} = P * (i / (1 - (1/(1+\text{i})^n)))$$

which in our case is

$$\text{ADSP} = \$1{,}000{,}000 * (0.08/(1-(1/(1+0.08)^5)))$$

or $250,456

To obtain the annual payment on Loan C, we use the same formula; only in our case, it is

$$\text{ADSP} = \$1{,}000{,}000 * (0.09/(1 - (1/(1 + 0.09)^{10})))$$

So, the annual payment on Loan C is $155,820.

Let us stop for a moment and compare* these three loans using the annual payment method of comparison.

Loan A	$280,000[a]
Loan B	$250,456
Loan C	$155,820

[a] Please note that this is just the first year's payment. As you know, with the level principal payment method, the annual payment reduces each year.

Is there any doubt in your mind as to which loan you will choose? There is no impeachment procedure that I know of for water authority board members; however, if you voted for any loan other than Loan C, I am certain that at least some of your ratepayers would try to impeach you or at least call for your resignation.

When you vote for Loan C, you are making a political decision. Nothing wrong with that. It is certainly your duty to contain the costs that your ratepayers must pay. However, what would be wrong is if you made that decision in ignorance of the true cost of these loans? To learn this, we must compare them by the total cost method, and to do this, we need to discount each of their annual payments. To do this, we will use a discount rate of 3% and use the discounting formula: $PV = FV / (1 + r)^n$. Here, of course, "FV" is the annual payment in any given year, "r" is the interest rate (0.08), and "n" is "the year of each respective annual payment." So, we have to apply the formula five times, once for each ADSP.

Here are the discounted values of the annual payments for Loan A:

	ADSP ($)	PV ($)
Year 1	280,000	271,845
Year 2	264,000	248,845
Year 3	248,000	226,925
Year 4	232,000	206,129
Year 5	216,000	186,323
Total[a]		**1,140,067**

[a] Adding up the ADSP column is nonsense. They are all apples and oranges.

* You can compare loans of only the same principal amount, but you can do so regardless of differing interest rates or differing terms. However, you cannot compare a $10,000 loan with a $12,000 loan.

Now, we can discount the annual payments on Loan B, as follows:

	ADSP ($)	PV ($)
Year 1	250,456	243,161
Year 2	250,456	236,079
Year 3	250,456	229,203
Year 4	250,456	222,527
Year 5	250,456	216,046
Total		**1,147,016**

Now, we will obtain the present values for each of the ADSPs in Loan C, as follows:

	ADSP ($)	PV ($)
Year 1	155,820	151,282
Year 2	155,820	146,875
Year 3	155,820	142,597
Year 4	155,820	138,444
Year 5	155,820	134,412
Year 6	155,820	130,497
Year 7	155,820	126,696
Year 8	155,820	123,006
Year 9	155,820	119,423
Year 10	155,820	115,945
Total		**1,329,177**

So, by comparing these three loans by the total payment method, we have the following:

Loan A	$1,140,067
Loan B	$1,147,016
Loan C	$1,329,177

As you can see, Loan A, which is the most expensive loan, when compared using the annual payment method, is the least expensive loan when compared using the total payment method. Conversely, Loan C, which is, by far, the least expensive loan when compared by the annual payment method, turns out to be the most expensive loan, by far, when compared using the total payment method.

So, as a board member of the water authority, you can certainly vote for Loan C, because it will cost your ratepayers the least, even though they may have to pay more over time.

Let us look at another example in which the inflation rate is in a state of flux, as it has been in the United States since 2008.

Let us say that an environmental utility manager is facing a 4%–6% rate of inflation and wants to measure his total project costs at two different rates, for comparison purposes. Let us say he has the following options for a $1,000,000 project:

Option A: 5% with a 5-year term (no points/fees)*
Option B: 6% with a 10-year term (no points/fees)

Both loans are level payment loans.

The level ADSP for Option A can be obtained by using our formula with the appropriate variables:

$$\text{ADSP} = \$1{,}000{,}000 * (0.05/(1 - (1/(1 + 0.05)^5))).$$

The level annual debt service for this loan is $230,975.

We can use the same formula to determine the level ADSP for Option B: $\text{ADSP} = \$1{,}000{,}000 * (0.06 / (1 - (1 / (1 + 0.06)^{10})))$, which is $135,868.

So, if we were only interested in comparing these two loans by the annual payment method, then we would have:

Option A	$230,975
Option B	$135,868

This looks like a no-brainer for our environmental utility executive, but there is more to it.

As we said above, our executive is confronted by a range of inflationary factors between 4% and 6%. So, to be sure, he needs to discount each of the annual payments at both 4% and 6%.

Here is Option A discounted at 4%.

	ADSP ($)	PV ($)
Year 1	230,975	222,091
Year 2	230,975	213,549
Year 3	230,975	205,336
Year 4	230,975	197,438
Year 5	230,975	189,845
Total		**1,028,259**

Now, here is Option A discounted at 6%.

	ADSP ($)	PV ($)
Year 1	230,975	217,901
Year 2	230,975	205,567
Year 3	230,975	193,931

(*Continued*)

* Points and fees will be taken up below.

	ADSP ($)	PV ($)
Year 4	230,975	182,954
Year 5	230,975	172,598
Total		**972,951**[a]

[a] Note that the PV of the annual payments of Option A is lower than the original principal amount of the loan. This is because the interest rate on the loan is below the rate of inflation that the utility executive is using to determine his true cost. This means that the bank must be using 3%, or maybe lower, as the rate of inflation that they see. Otherwise, a banker will soon be fired or his bank will close its doors for good.

Now, let us look at Option B. First, let us discount loan B's annual payments at 4%.

	ADSP ($)	PV ($)
Year 1	135,868	130,642
Year 2	135,868	125,618
Year 3	135,868	120,786
Year 4	135,868	116,141
Year 5	135,868	111,674
Year 6	135,868	107,378
Year 7	135,868	103,249
Year 8	135,868	99,277
Year 9	135,868	95,459
Year 10	135,868	91,788
Total		**1,102,012**

Now, let us look at Option B's level annual payments discounted at a rate of 6%.

	ADSP ($)	PV ($)
Year 1	135,868	128,177
Year 2	135,868	120,922
Year 3	135,868	114,077
Year 4	135,868	107,620
Year 5	135,868	101,528
Year 6	135,868	95,782
Year 7	135,868	90,360
Year 8	135,868	85,245
Year 9	135,868	80,420
Year 10	135,868	75,868
Total		**1,000,000**[a]

[a] For those of you checking the work, I was rounding. The numbers actually add up to $999,999. When the interest rate on a loan equals the particular discount rate that you are using, the answer is always the original principal amount of the loan.

So, now, here is what both Option A and Option B look like under both inflation rate scenarios:

Inflation Rates	**4%**	**6%**
Option A	$1,028,259	$972,951
Option B	$1,102,012	$1,000,000

As you can see, in each instance, Option A offers the lowest cost to the utility executive, regardless of either discount rate. It would be especially good for his utility if his costs (and his revenues) were increasing at 6%. Then, the loan would actually make the system a modest profit, odd as that may sound.

Please note that the higher the discount rate, the lower the present value. (This works in reverse for compounding. The higher the compounding rate, the higher the future value.)

It is very tempting simply to add up all the annual payments for different loans and then compare them, but this is incorrect. The correct way to compare loans is to discount the future payments to their present value. The present value is their value in today's dollars. You must always compare loans in today's dollars.

Here is the brief methodology for comparing loans.

Method Rules

1. Write out the annual payment schedule, or amortization schedule, of each loan you are considering, including the amount in dollars that must be paid in each year along with the year in which it must be paid.
2. Choose the discount rate that you will use.
3. Convert each annual payment to its present value by discounting it for the number of years in the future in which it must be paid.
4. Sum up all the annual present values for each financial option that you are considering.
5. Compare the present values of each alternative.

Points, Fees, and True Interest Cost

It seems like it is totally impossible to borrow money without having to pay one or more fees or charges. That said, we need to know what impact fees or charges will have on the ADSPs that ratepayers must pay. This will lead us to the important concept of true interest cost (TIC) or what the banks euphemistically refer to as the annual percentage rate (APR).

There are three types of fees or charges:

1. One-time-only, upfront fees, such as the "points" the bank charges you when you take out your mortgage. This also includes other one-time charges such as attorney's fees, appraisal fees, and the like.

2. Annual fees that are expressed in flat dollar amounts. These are charges such as a $25 per month "servicing fee."
3. Annual fees that are expressed as a percentage of the outstanding principal balance. Some monitoring fees are expressed in this way. For example, on a municipal bond, the bank, acting as bond trustee, might charge an annual monitoring fee of 1/8th of one per cent (0.125%). (This actually happens, but it is a wonder why. Over time, the outstanding principal balance of the loan declines and so does the 0.125% fee. However, the cost of monitoring [personnel, facilities, etc.] increases. So, you have the dire circumstance of declining income and rising expenses.)

Each of these types of charges is dealt with differently. Here is how they work:

1. One-time, upfront fees are added to loan principal. So, when your bank charges you two points on your $200,000 mortgage, or $4,000, you actually have a $204,000 loan, not a $200,000 loan.
2. Annual fees are added to each annual payment. In this case, by adding a $25 a month fee to your $1113 monthly payment, it becomes a $1128 monthly payment.
3. Annual fees that are expressed as a percentage of the outstanding principal balance are added to the interest rate. Here, your 6% loan with a 1/8th per cent monitoring fee becomes a 6.125% interest rate.

This is all well and good, but what impact do these charges have on your monthly or annual payments?

Finding the impact of one-time, upfront charges is very difficult without a financial calculator. Moreover, then it is not really easy.

Essentially, we use our favorite formula for calculating ADSPs, $ADSP = PV * (i / (1 - (1 / (1 + i)^n)))$, but we have to calculate backward to find "i", the interest rate. In other words, the bank's adding points have the same impact on your loan as if the bank raised the interest rate. The bank could, of course, just raise the interest rate, but then, you might decline the loan. In addition, the bank would get its money over 30 years, but with points upfront, it gets the extra money today.

If your bank charges you two points on your $200,000 mortgage and charges you a 5% rate of interest, here is what happens.

As we said before, you add the points to principal. So, you now have a $204,000 mortgage at 5% for, say, 15 years. In such case, your annual* payments would be $19,653.83. Now, if the bank did not charge you the two points (or you paid it in cash), you would have a $200,000 mortgage at 5% for 15 years. As such, your annual payment would be $19,268.46. So, what does this mean to your effective borrowing rate?

To calculate this effect, we take the ADSP on the $204,000 loan, but then take $200,000 as the loan principal, all at the 15-year term. This complicated effort

* Even though mortgages have monthly payments that are slightly different than simply multiplying them by 12 to get an annual payment, we are going to use a simple annual payment (like most public debt) for our examples here.

reveals that your $200,000 mortgage with two points is the equivalent of having a 5.29% mortgage rate. When you finalize the paperwork for this loan, one of the little documents the bank will give you is a notice saying that, although the nominal rate of interest on your loan is 5%, your equivalent rate (because of the two points) is 5.29%. This is what the bankers call the APR. Not too odious a euphemism. However, the rest of the finance industry calls this phenomenon the TIC, words that are much closer to the plain truth.

Annual flat fees are dealt with, as we said, by adding them to the annual payment. Let us look at a $100,000 loan for 20 years at a rate of 5%. The level annual payment on this loan would be $8,024.26. Adding $25 a month, or $300 a year, to this payment would make it $8,324.26. Now, what impact does this $25 per month have on your TIC?

This time, we take $8,324.26 as our annual payment, 20 years as our term, and $100,000 as our principal. Moreover, we calculate a new interest rate, which will be our TIC. In this case, it is 5.44%. So, in this case, our nominal rate of interest is 5%, but our TIC is 5.44%.

The last type of charge, an annual fee expressed as a percentage of the outstanding principal balance, is the easiest to deal with. All you do is add it to your interest rate. So, if your nominal interest rate is 5% and you are being charged 1/8th of 1% (0.125%) monitoring fee, then your TIC is 5.125%.

Comparing loans or bonds, both with and without fees and charges, is essential to identifying the lowest cost alternative for ratepayers or the general public to pay. In addition, identifying the lowest cost alternative is essential to providing the greatest environmental benefit to the largest possible number of people.

6 Rate Setting

Efficient rates reduce the cost of providing environmental utility services to customers.

The following are some characteristics of well-considered rates for water and wastewater systems:

- *Revenue sufficiency/cost recovery*: Rates produce stable revenue that is equal to the financial cost of supplying the water or wastewater service.
- *Fairness*: Equals must be treated equally. In other words, the prices charged to customers are equal to the costs imposed on the system by those customers.
- *Resource conservation*: Pricing decisions should not promote the unwise use of water resources.
- *Net revenue stability*: Prices should allow the utility to have sufficient income to meet its operating costs, even when quantities demanded are below a normal level.
- *Transparency*: Pricing structures should be able to be understood by every consumer, in order for the consumer to respond accurately when deciding how much water to consume.
- *Ease of implementation*: The pricing structure should not impose significant administrative costs on the utility.
- *Affordability*: The prices charged to customers should be within an affordable range.

The objectives listed above provide a guideline for utility managers to follow when deciding on water rates; however, not all of the objectives can be met at the same time. This chapter will focus on the objective that is most directly related to the financial stability of the utility: cost recovery. The objectives of equity and affordability will be discussed in Chapter 7.

FULL COST RECOVERY RATES

When setting the rate, utility managers must understand that they must at least be able to recover the cost of operating the water system from the revenues earned. The rate must be set at a level where it will equal the amount of total cash expenses. One method for designing a rate is to begin with the operating budget that shows total cash expenses. Next is to devise a "residential equivalent" (RE) to measure commercial usage. (You will also see the concept of a RE in the discussion on stormwater in Chapter 16.) For example, an office building might use water and sewer services equal to 10 average households. Next is to divide the total cash expenses by the total number of households—plus the total number of REs—that will purchase the service; the result is the average residential rate required to recoup the costs of operating the system.

IMPACT OF FINANCING PROJECTS

It is highly unlikely that a utility can rely on grant funding for 100% of its project costs; it is probable that the utility will have to pay at least a portion of the project costs. The more project costs that are paid for by the utility, the higher the household water rates, since the utility must pass on the costs of providing services to the customers that use water. A helpful exercise will be to determine what must be the rate for a community that is in need of a project to modernize its water system, so that the full cost of water delivery is recouped by the utility. This exercise will be considered for 4 different scenarios: 100% of the project funded by an outside grant; 75% of the project funded by a grant and 25% funded by a loan; 50% of the project funded by a grant and 50% funded by a loan; and 25% of the project funded by a grant and 75% funded by a loan. Under these different scenarios, the important factors will be the term of the loan and the interest rate. Since income must equal the expenses (operating plus annual project costs), it is important to consider the effect of rate and term on the household rate, so that the rate does not become unaffordable.

In the following example, assume that your utility serves 1000 households and your current expenses* of $10,000 per year are covered completely by a $10 per household rate. In order to modernize the water system and to repair pressing problems, a project is proposed that will cost $1500 plus an increase in annual operating expenses of $550. Below are a few possible effects on the rate corresponding to different project-financing options.

Financing Option	Annual Operating Cost ($)	Annual Project Cost[a] ($)	Total Annual Cost ($)	Full Cost Recovery Rate ($ per household)	Percentage Change in Rate
100% grant	10,550	0	10,550	10.55	5.5
75% grant, 25% loan (5 years, 10%)	10,550	112.5	10,662.50	10.66	6.63
75% grant, 25% loan (10 years, 5%)	10,550	56.25	10,606.25	10.61	6.06
50% grant, 50% loan (5 years, 10%)	10,550	225	10,775	10.78	7.75
50% grant, 50% loan (10 years, 5%)	10,550	112.50	10,662.50	10.66	6.63
25% grant, 75% loan (5 years, 10%)	10,550	337.50	10,887.50	10.89	8.88
25% grant, 75% loan (10 years, 5%)	10,550	168.75	10,718.75	10.72	7.19
100% loan (5 years, 10%)	10,550	450	11,000	11.00	10.0
100% loan (10 years, 5%)	10,550	225	10,775	10.78	7.75

[a] Using a level principle payment schedule, the total annual payment decreases annually; in this example, the annual project cost is equal to the first year's payment; therefore, subsequent year payments will be lower.

* Expenses include energy, labor, chemicals, and administration costs.

As expected, the more a utility must pay for the project costs in a full cost recovery rate system, the larger the impact on the household rate. As shown above, a longer-term loan has a lesser impact on the costs (and therefore rate) than a shorter-term loan.

RATE DESIGN OPTIONS

Given the requirement that utilities use full cost recovery methods, as described above, there are a variety of ways in which a utility can design the rate scheme. To demonstrate the unique features of each rate design option, the following example will be applied to each option.

A public utility has 100 customers, who purchase the following amounts of water each month:

- 25 customers consume 500 m^3 of water
- 50 customers consume 1000 m^3 of water
- 25 customers consume 1500 m^3 of water

FIXED VERSUS VOLUMETRIC CHARGES

Fixed Charges

The first way in which we will distinguish rates is by those that have a fixed charge and those that have a volumetric charge.

A fixed charge is one where the amount of money charged to a customer is the same each month and is independent of the amount of water that the customer consumes.

In places where there are no water meters, volumetric charges are not a viable option, because there is no way to accurately measure how much water each customer consumes. In this situation, a "single-part rate," or "fixed charge rate," using equal monthly charges, is the only available option.

When setting the fixed charge, or fee, for all customers, the utility must first determine its own costs for delivering the amount of water demanded by all customers, because the total revenue that the utility will receive must cover the utility's total costs and also provide funds for maintenance and repairs. A utility may define different "classes" of customers, so that residential customers are charged different (usually lower) fixed rates than business customers, since businesses tend to consume more water than residential users.

Fixed charges, when used alone, have two major drawbacks. First, consumers are given no incentive to economize on water use, since each additional gallon of water comes free of charge. Second, as use of water increases, which will certainly happen with a growing population and a developing economy, the utility's ability to recover its costs by the fixed charge will diminish, because the costs of meeting growing needs will increase.

In our example, all 100 customers are charged the same fee (in this case, it is set at $20 each month), even though they consume different amounts of water.

Consumption Level (cm)	Total Charge ($)
500	20
1000	20
1500	20

Volumetric Charges

A utility that has water meters on all or most household connections is able to monitor the amount of water that each consumer (i.e., household) uses and is thus able to charge customers according to the amount of water consumed each month. When setting this price, the utility should attempt to set the volumetric charge at the "marginal cost" per unit of water provided to each consumer.

The marginal cost estimate must incorporate administrative costs as well as the costs associated with each additional volume of water provided, because, in a single-part rate, there is no additional fixed charge added to consumers' bills to cover the administrative costs.

To calculate the total charge for each customer group (in this case, groups are separated by the amount of water consumed), multiply the amount of water consumed by the price per gallon of water. In this case, we have a uniform volumetric rate structure, where the price per unit of water is the same regardless of how much water is purchased in total by each customer.

In this example, a "uniform volumetric charge" is used, where each consumer group is charged the same volumetric price. The next section discusses other forms of volumetric charges.

The volumetric charge example is different from the fixed charge example in that the charges now depend on the amount of water consumed, and therefore, each customer group has a different total bill for water. The total charges are calculated as follows, using a uniform volumetric charge of $0.02 per cubic meter*:

Consumption Level (cm)	Total Charge Calculation	Total Charge ($)
500	500 × $0.02	10
1000	1000 × $0.02	20
1500	1500 × $0.02	30

MULTIPART RATE

A rate can be made up of either a fixed or a volumetric charge, as explained above, or it can have both, which would be a multipart rate. In the case of a multipart rate, the fixed charge element of the water bill tends to cover administrative costs (e.g., the cost of installing and reading water meters and billing customers) incurred by the utility in order for it to be able to deliver water to each customer, and the volumetric charge amount reflects the "per unit" cost of delivering water to the consumer.

* A cubic meter is about 268 gallons.

The fixed charge is in addition to the price of delivering the amount of water demanded by that customer (i.e., the volumetric charge), because it reflects the costs incurred by the water utility to provide services to customers, regardless of the level of water consumption.

In our example, the fixed charge will be \$10 per connection (i.e., household) and the uniform volumetric charge will be \$0.02 per cubic foot of water consumed.

Consumption Level (cm)	Total Charge Calculation (\$)	Total Charge (\$)
500	\$10 + (500 × \$0.02)	20
1000	\$10 + (1000 × \$0.02)	30
1500	\$10 + (1500 × \$0.02)	40

Types of Volumetric Charges

There are a small number of accepted methods for designing volumetric charges that must be considered when setting rates, so that the utility is able to select a rate that will achieve full recovery of the costs to deliver water service, while trying to achieve the other goals of equity, affordability, and economic efficiency. The four volumetric pricing design options, in addition to the uniform volumetric charge that has been described above, are increasing block rate (IBR), declining block rate (DBR), seasonal pricing, and zonal pricing.

Increasing Block Rate

In theory, IBR can achieve three objectives simultaneously

1. Promote affordability by providing the poor with affordable access to a "subsistence block" of water
2. Achieve efficiency by confronting consumers in the highest-price block with the marginal cost of using water
3. Raise sufficient revenues to recover cost

However, in practice, IBRs often fail to meet any of the three objectives mentioned above, in part because they tend to be poorly designed. Many IBRs fail to reach cost recovery and economic efficiency objectives, usually because the upper consumption blocks are not priced at sufficiently high levels and/or because the first subsidized block is so large that almost all residential consumers never consume beyond this level.

When used in a multipart rate, the increasing block pricing scheme affects only the volumetric part of the total rate; the fixed charge remains the same for all customer groups. Calculating the total charge for each customer group by using an increasing (or a declining) block rate first requires delineating the number of blocks and the quantity of water allowed in each block. Customers will pay the fixed charge plus the sum of the products of the amount of water consumed in each block multiplied by the per unit price of water in each block.

In our example, the three blocks will be as follows:

- Block 1: $0.01 per cubic meter for first 500 cm of water consumed
- Block 2: $0.02 per cubic meter for amounts between 501 cm and 1000 cm of water consumed
- Block 3: $0.03 per cubic meter for amounts above 1000 cm of water consumed

Consumption Level (cm)	Total Charge Calculation	Total Charge ($)
500	$10 + (500 × $0.01)	15
1000	$10 + [(500 × $0.01) + (500 × $0.02)]	25
1500	$10 + [(500 × $0.01) + (500 × $0.02) + (500 × $0.03)]	40

Notice how the incremental increase of 500 cm of water resulted in an exponentially higher total charge for the third block than for the second block.

The next rate design option to be discussed is the DBR.

Declining Block Rate

A DBR is the opposite of an IBR. With a DBR, consumers face a high volumetric charge up to the specified quantity of the first block. Then, any water consumed beyond this level, and up to the next block, is charged at a lower rate, and so on for as many blocks as the rate utilizes.

The DBR structure was designed to reflect the fact that when raw water supplies are abundant, large industrial customers often impose lower average costs because they enable the utility to capture economies of scale in water source development, transmission, and treatment. This rate design has gradually fallen out of favor, in part because marginal costs, properly defined, are now relatively high in many parts of the world, and thus, there is increased interest in promoting water conservation by the largest customers. The DBR structure is also often politically unattractive because it results in high-volume users paying lower average water prices.

In the DBR example, we will continue to use a fixed rate of $10 and the block delineations; however, the per unit price of water is now

- Block 1: $0.03 per cubic meter for first 500 cm of water consumed
- Block 2: $0.02 per cubic meter for amounts between 501 cm and 1000 cm of water consumed
- Block 3: $0.01 per cubic meter for amounts above 1000 cm of water consumed

Consumption Level (cm)	Total Charge Calculation	Total Charge ($)
500	$10 + (500 × $0.03)	25
1000	$10 + [(500 × $0.03) + (500 × $0.02)]	35
1500	$10 + [(500 × $0.03) + (500 × $0.02) + (500 × $0.01)]	40

Notice how the total charge for the first two groups' users is now $10 higher than that with the IBR, whereas the total charge for the largest water user did not change.

Seasonal Pricing

In some circumstances, the marginal cost of supplying water to customers may vary by seasons. In such cases, water rates can be used to signal customers that the costs of water supply are not constant across the seasons. Water use in summer tends to be much higher for households that have gardens and any other high water uses; the increase in water use is usually found to be in outdoor water use, whereas indoor water use tends to remain constant throughout the year.

When the marginal cost to provide water services changes according to seasons, then utilities can charge higher prices during the more costly seasons (usually summer) and lower prices during the less expensive seasons (usually winter).

In this example, we will return to using the uniform volumetric rate, which charges the same per unit amount for water, regardless of the amount of water consumed, so that we can focus on the effect of a higher price for the summer season than for the winter season. We will assume that the amount of water consumed at all levels is constant during the year. The fixed charge is $10 and the seasonal prices are as follows:

- Winter: $0.01 per cubic meter of water consumed
- Summer: $0.02 per cubic meter of water consumed

Consumption Level—Summer (cm)	Total Charge Calculation	Total Charge ($)
500	$10 + (500 × $0.02)	20
1000	$10 + (1000 × $0.02)	30
1500	$10 + (1500 × $0.02)	40

Consumption Level—Winter (cm)	Total Charge Calculation	Total Charge ($)
500	$10 + (500 × $0.01)	15
1000	$10 + (1000 × $0.01)	20
1500	$10 + (1500 × $0.01)	25

A seasonal and/or zonal pricing (described below) scheme(s) can be added to any of the three pricing methods above (uniform, IBR, and DBR), as they are simply adjustments to the chosen pricing scheme, based on the time of the year (seasonal pricing) and the location of customers (zonal pricing).

Zonal Pricing

Zonal Pricing can occur when the marginal cost to provide water service varies according to the location of the customer. In this case, utilities can charge higher volumetric rates to the customers that live in an area that is more costly to serve and lower volumetric rates to those that live in the areas that are cheaper to serve.

It may cost the water utility more to deliver water to outlying communities due, for example, to higher elevations and increased pumping costs. Zonal prices can be used to ensure that users receive the economic signal that living in such areas involves substantially higher water supply costs. However, this type of special rate is appropriate only if the costs to serve the area are significantly higher than for the rest of the community—in fact, costs vary among all users, and a practical rate always reflects average costs to some degree.

In the zonal pricing example, we will continue to use a fixed charge of $10 and a uniform volumetric charge with variations in the per unit charge, depending on the zone in which consumers reside.

- Zone 1: $0.02 per cubic meter of water consumed
- Zone 2: $0.03 per cubic meter of water consumed

Consumption Level—Zone 1 (cm)	Total Charge Calculation	Total Charge ($)
500	$10 + (500 × $0.02)	20
1000	$10 + (1000 × $0.02)	30
1500	$10 + (1500 × $0.02)	40

Consumption Level—Zone 2 (cm)	Total Charge Calculation	Total Charge ($)
500	$10 + (500 × $0.03)	25
1000	$10 + (1000 × $0.03)	40
1500	$10 + (1500 × $0.03)	55

The goals of full cost recovery and economic efficiency are achieved when utilities charge customers according to the cost that each customer imposes on the system. If the cost to deliver water is the same for all of the utility's customers, then a uniform volumetric charge, which is set equal to the cost of water delivery, is the most efficient. The DBR is used when the cost to deliver large amounts of water is cheaper than that for small amounts of water. When the marginal cost of water is high, IBR is usually chosen over DBR, since IBR charges higher rates to customers that demand higher levels of water.

7 Subsidies and Affordability

A year or so after the Clean Water Act was amended in 1987 to create the Clean Water State Revolving Fund (CWSRF), some of the thoughtful souls at the US Environmental Protection Agency (EPA) began to wonder if communities could afford the new loan program. Grants were free, but loans are not. Were communities across the country going to be able to repay these new loans?

So, they convened a meeting at EPA for the specific purpose of discussing the "concept of affordability." Most people who attended the meeting were members of EPA, but a few outsiders, like us, were also invited. These discussions went very well, but there was one gentleman who sat there with his arms folded across his chest, squirming in frustration. Finally, no longer able to contain himself, he blurted out, "In the world of classical economics, there is no such thing as the 'concept of affordability,'" to which another of the participants gently responded: "Yes, but in the real world, there is no such thing as classical economics."

There are two affordability issues: Community affordability and individual affordability. They are not mutually exclusive. You will have cases of individual unaffordability in the wealthiest communities, and, perhaps surprisingly, you will also have cases of individual unaffordability in places where there is community unaffordability.

We will deal with community affordability first.

COMMUNITY AFFORDABILITY

Back in the mid-nineties, the Environmental Finance Center at the University of Maryland convened a meeting to discuss the dilemma of a small town in Western Maryland, situated on a tributary of the Potomac River. Just a few feet below the surface of the land was a stratum of solid bedrock. This was a rural community, with all homes built on septic systems. However, because of the bedrock below, the septic systems did not work very well. The scientists from the state were puzzled with the readings they were getting on their downstream water quality tests. After much testing and investigation, they concluded that the problem was coming from all of the town's septic systems on the bedrock. So, the state ordered the town to install sewers. There were two problems with this requirement. First, to install the sewer mains, they would have to blast out the bedrock. Second, the homes in this town were not very close together. They were going to need several hundred extra feet of sewer mains. So, this was going to be a very expensive enterprise.

This was not a wealthy town to begin with. However, back in those days, the average home value was under $100,000. The residents, then, were understandably shocked when the engineers concluded that the project would cost $39,000 per home!

This is a classic case of where a grant is needed. It is a classic case of community *un*affordability. (And it is not classical economics!)

There is a rule of thumb in the environmental infrastructure business that says that rates of 1% of median household income (MHI) for drinking water and 1% for wastewater are affordable, which means 2% for water and sewer combined. Notwithstanding the general consensus about the 1% each affordability number for water and sewer service, EPA thinks that it should be 2% each. Regardless of what EPA thinks, most people think 1% for each is a good number, so that is what we will use.

As a matter of fact, the American Water Works Association in partnership with Raftelis Financial Consultants, Inc. undertook their "2010 Water and Wastewater Survey" of 49 states and the District of Columbia. With 308 responses for water and 288 for wastewater, the report found that the average water charge was 0.66% of MHI and the average wastewater charge was 0.84% of MHI. The total then is 1.5% for both. So, our 2% is a conservative number.

It is a conservative number in one sense: that of "measuring a community's, not an individual's, ability to pay."

Good old, tax-exempt municipal bonds are a form of subsidy. Systems get to pay lower interest rates, because the income is not taxable to the bondholder. Furthermore, all of the state CWSRF programs offer generous subsidies. Most lend at approximately 50% of the market rate. So, when AAA tax-exempt municipal bonds are selling at 4%, the most CWSRFs are making loans at about 2%. Does anybody think that any environmental utility is going to give up these subsidies? No, of course not.

The failure to set full cost rates is frequently attributed to community affordability limits, which may be wrongly characterized in terms of the community's ability to pay. In fact, the underlying cause is generally the existence of relatively large numbers of households with individual affordability problems. This creates a political climate in which significant increases in rate level are seen as unacceptable, owing to the harm that would be inflicted on the lowest-income customers. (This situation can, of course, be effectively dealt with by targeting subsidies to the households with true need.)

However, in the absence of unlimited outside subsidies, utilities generally deal with this situation by trying to reduce costs through such ill-advised strategies as deferring maintenance, deferring facility upgrades and replacements, eliminating staff functions, and maintaining low wage levels. One of the most common manifestations of these funding problems in older American cities is a deteriorated water or sewer system that is plagued by leaks and overflows.

INDIVIDUAL AFFORDABILITY

To address the question of individual affordability, we need to identify all those families near the bottom of the income curve that cannot afford a service and to set up a subsidy for only those poor households.

Even in the richest county in the country, Loudoun County, Virginia, with its $119,000 MHI, there are undoubtedly some families that cannot pay their water and sewer bills. In most cases, when we stop the subsidies for the many who do not need them, we will have more than enough money to subsidize the few that truly do.

That said, the first step is to identify those who can't pay their bills. There are several ways of doing this. The most straightforward is to tell customers that subsidies are available to households with less than $X annual income and then ask the customers wishing subsidies to show their last one or two federal income tax forms. Another way is to contact the state agency handling food stamps and/or welfare payments. Yet another way is to find out which customers qualify for the federal Low Income Household Energy Assistance Program (LIHEAP), which helps the poor pay their energy bills. (If there are confidentiality issues that prohibit obtaining the names of customers on food stamps, welfare, LIHEAP, and so on, then the customers can again be asked to step forward voluntarily.)

Now, having identified those who cannot afford their utility service, we now need to address this individual affordability issue by identifying appropriate subsidies to help them pay their bills.

There are at least three choices here.

1. The subsidy can come from general tax revenues, through some established state or local program. As you will see in our example below, we have identified 4,000 households (of 100,000) that cannot afford their utility rates. They each need a subsidy of about $20 per month, or $240 per year, for a total of $480,000. In this first case, the subsidy would come from the state or local government, or both, out of their general revenues. In other words, the local or state government would take $480,000 out of their general fund and pay it to the utility.
2. The subsidy can be a cross-subsidy. In this case, the ratepayers would be divided into two categories: hardship and nonhardship. In this case, the subsidy for the hardship ratepayers comes from the nonhardship ratepayers. This method is illustrated below.
3. Finally, the subsidy can come from all utility customers, whether or not they can be considered hardship customers.

Each of these options has advantages and disadvantages. The first option—subsidy from state or local tax revenues—may not be possible or feasible.

The difference between the last two options can be illustrated by a very simple example. This example considers only the case where there is no outside subsidy available to the utility—the program is funded by increasing some charges to some, or all, customers. The example is constructed for a medium-sized community of 100,000 households. We will assume that the MHI for this community is about $50,000, which is about the national MHI. We will also assume that the MHI for our hardship cases is about $20,000, which is below the poverty level for a family of four.

Using our rule-of-thumb affordability number of 1% of MHI, the community in general should be paying about $500 a year. In our example, they will pay $40 per month, which is $480 yearly. Close enough.

Our hardship cases should be paying only $200 per year. However, in our case, we are going to target a subsidy of $20 per month for them. This means, of course, that they should be paying $20 per month or $240 per year. This is just above 1% of MHI, actually 1.2% of MHI, but so be it.

As you will see below, our system has 100,000 ratepayers, of whom 4,000 are hardship cases. So, to create a $20 per month subsidy for the hardship cases, we need a subsidy of $40,000 per month or $480,000 per year.

Assume:

Number of (residential) customers = 100,000
Number of hardship customers = 4,000
Total water sales = 750 MG/month
Price of water = $2 per 1,000 gal
Average water sales per household = 7,500 gal

1. *Base case:* The original, subsidy-free rate design. Hardship customers pay exactly the same bill as do all other ratepayers.
2. *Targeted subsidy, financed by nonhardship customers:* This design incorporates a subsidy of $20.00 per month for only hardship customers, financed by increasing the fixed charge for nonhardship customers. Note that the increased cost for nonhardship customers is small.
3. *Targeted subsidy, financed by all customers:* This is the same as design B, except that all customers contribute equally to financing the subsidy. The results (as compared with B) are a slightly lower cost for nonhardship customers and a slightly larger payment (as compared with B) for hardship customers.

	A	B	C
Nonhardship Customers			
Fixed charge ($/month)	25.00	26.22	25.40
Variable charge ($/1,000 gal)	2.00	2.00	2.00
Monthly variable charge	15.00	15.00	15.00
Average bill ($/month)	40.00	41.22	40.40
Hardship Customers			
Fixed charge ($/month)	25.00	5.00	5.40
Variable charge ($/1,000 gal)	2.00	2.00	2.00
Monthly variable charge	15.00	15.00	15.00
Average bill ($/month)	40.00	20.00	20.40

As you can see, the impact of both alternatives B and C for the nonhardship customers is totally minimal. Likewise, the subsidy is extremely valuable to the hardship customers. Even in alternative C, the hardship customers' monthly bills are only 2% higher than in those in alternative B.

Whether you believe in classical economics or not, affordability is a real issue that needs to be dealt with. It is one of the few legitimate uses of subsidies.

8 The Impact of Term on Annual Debt Service Payments

We must begin this analysis by saying that the term of a loan has an enormous impact on annual debt service payments. This is true whether you use the level payment (LP) method or the level principal payment (LPP) method.

As you will see, the impact of term on annual debt service makes the decision of whether to choose a loan with the LP method or the LPP method a critical judgment. For this reason, we will discuss the impact of term on annual debt service separately for each type of payment method. Then we will compare the two types of loan payment methods in terms of their respective impacts on annual debt service payment.

LEVEL PRINCIPAL PAYMENT LOANS

If the term of a loan is only 1 year, then 100% of the principal is due and payable at the end of that year. If the term of an LPP loan is 2 years, then 50% of the principal balance is due and payable at the end of each year. For a $100,000 loan at an interest rate of 7%, the annual debt service payment at the end of the first year would be $57,000, consisting of $50,000, or 50%, of the principal and $7,000 of interest, representing 7% on the $100,000 of principal, which was outstanding during the first year of the loan.

However, the second annual debt service payment, which would be due at the end of Year 2, would only be $53,500. This payment is composed of $50,000 of principal, which is the remaining principal outstanding, and $3,500 of interest, representing 7% on the $50,000 of principal, which was outstanding during the second year of the loan.

Therefore, the annual debt service payment at the end of Year 1 would be $57,000 and at the end of Year 2 would be $53,500. The difference between these two payments is only $3,500, which is about 7%. A 7% difference in annual debt service payments cannot be considered significant.

However, when we start talking about loans with terms of more than 2 years, the story begins to change radically.

Let us use the same $100,000 loan and the same 7% interest rate, but this time, let us use a 20-year term. In doing so, we can examine the difference between the first and the last annual debt service payments.

Assuming, once again, that we are dealing with an LPP loan, the amount of principal, which would be repaid every year, would be 1/20th of $100,000, or $5,000. For the first year, the outstanding principal balance would be the original principal balance, or $100,000. The interest on $100,000 for 1 year at a rate of 7% would be $7,000. Since the principal payment at the end of the first year would be $5,000 and the interest payment would be $7,000, the total annual debt service payment for Year 1 would be $12,000.

For Year 20, the outstanding principal balance would only be $5,000, since the other $95,000 had been repaid in the preceding 19 years. The interest, at a rate of 7%, on the outstanding principal balance of $5,000 would be $350. Therefore, the total annual debt service payment for Year 20 would be $350 of interest and $5,000 of principal, or $5,350.

Now, compare the first year's annual debt service payment of $12,000 with the annual debt service payment in Year 20 of $5,350. Here, the difference in the two annual debt service payments is $6,650 or a 55% decrease! I am certain that the ratepayers in Year 20 would be pleased with this circumstance, but what about the ratepayers in Year 1, who are getting the benefit of the same project? That's the problem with LPP loans.

LEVEL PAYMENT LOANS

Let's look at the same project with an LP loan.

As you know, an LP loan means that each annual debt service payment is equal to any other annual debt service payment. If the service life of, say, a new treatment facility is 20 years, then using our example described above of a $100,000 loan at an interest rate of 7%, the annual debt service payment under the LP method would be $9,439. This means that the annual debt service payment would be $9,439 in each of the 20 years the loan was outstanding. In such case, the total debt service payment would be 20 times $9,439 (20 × $9,439), or $188,780.

A ratepayer in Year 1 would pay his or her share of the $9,439, which is 1/20th of the total project cost. Assuming that the project—say a new treatment facility—had a service life of 20 years, that ratepayer would also enjoy 1/20th, or 5%, of the benefit of the project. In other words, in Year 1, he or she would pay 5% of the cost of the project and would receive 5% of the benefit of the project. Fair enough.

Now, the theme of this chapter concerns the effect of rate on term. Let us look at the same two loans if they had 40-year terms instead of 20-year terms.

On the LPP loan, the annual principal payment on a 40-year, $100,000 loan would be 1/40th, 2.5%, or $2,500. The interest payment would be the same in Year 1 as for the 20-year loan (since the full $100,000 is outstanding in the first year), or $7,000. So, the annual debt service payment in Year 1 for the LPP loan would be $9,500. For the 40th year, only $2,500 would remain outstanding. The interest on this amount, at the 7% rate, would be $175. In consequence, the 40th year annual debt service payment would be $2,675.

The difference between the annual debt service payment in Year 1 ($9,500) can be characterized as being 258% "higher" than the annual debt service payment in Year 40 ($2,675). How do you think your ratepayers would react to those numbers?

Now, look at the annual debt service payments of a 40-year LP loan. Here, the annual payment in each year is $7,501.

Note that the annual payment on the LP loan in Year 1, $7501, is significantly lower than the Year 1 payment on the LPP loan, $9,500. It is 21% lower, to be exact.

To put the concept of the power of term into full perspective when dealing with LPP loans, below is a table that sets forth the first years' annual payments on a $100 at several different interest rates over several different terms:

	Interest Rates				
Term	**0%**	**5%**	**10%**	**15%**	**20%**
1 year	$100	$105	$110	$115	$120
2 years	$50	$55	$60	$65	$70
3 years	$33	$38	$43	$48	$53
4 years	$25	$30	$35	$40	$45
5 years	$20	$25	$30	$35	$40
10 years	$10	$15	$20	$25	$30
20 years	$5	$10	$15	$20	$25
30 years	$3	$8	$13	$18	$23
40 years	$2.50	$7.50	$12.50	$17.50	$22.50

Please note the hugely significant impact that term has on the annual debt service payments for these loans.

Now, we turn to the LP method.

The impact of term on annual debt service payments is as significant when the LP method is used as when the LPP method is used. With the LP method, however, there is no problem with fairness. There is no problem with an adverse characterization of the numbers.

	Interest Rates				
Term	**0%**	**5%**	**10%**	**15%**	**20%**
1 year	$100	$105	$110	$115	$120
2 years	$50	$54	$58	$62	$65
3 years	$33	$37	$40	$44	$47
4 years	$25	$28	$32	$35	$39
5 years	$20	$23	$26	$30	$33
10 years	$10	$13	$16	$20	$24
20 years	$5	$8	$12	$16	$21
30 years	$3	$7	$11	$15	$20
40 years	$2.50	$6	$10	$15	$20

Again, you can see the enormous impact of term on annual debt service payments. The payment on a 10-year loan at 5% is $13, whereas it is $8 on a 20-year loan. This means that the 10-year loan costs ratepayers 63% more than does the 20-year loan. In addition, of course, if the service lives of the assets being financed were 20 years, why would any utility favor the shorter-term loan?

So, as we have said several times so far in this book, term is one of the most powerful concepts for lowering the cost of environmental improvement projects. In addition, as we know, the less expensive such projects are, the more will get done.

A few years ago, when California was beginning to organize their energy-efficiency loan program, I met an air quality officer in the Central Valley. She told me she had just taken out a second mortgage to install insulation in her house and to replace all doors and windows. She said that her bank gave her a 7% loan for 7 years. Assuming she borrowed $10,000, her monthly payment would have been $151.

Now, there is no reason—other than bank policy—why a home improvement loan for new doors, windows, and insulation should have only a 7-year term. Those improvements last as long as the house itself. Surely, a 20- to 30-year loan would have been more appropriate.

Now, this woman's county had not organized their energy-efficiency loan program under the new state law (see Chapter 18). However, if they had organized such a program, she could have gotten a 20-year term (without the second mortgage) loan. Even assuming the same 7% rate of interest, her monthly payment would have been $78!

Recall that the first principle of environmental finance is to create the greatest amount of environmental benefits for the largest number of people for the lowest cost. The second principle is that the more we can drive down the cost of environmental improvements, the more will get done. This is a golden example of these two principles. More people are likely to insulate their homes at a cost of $78 than at a cost of $151, and that is our goal!

Term is one of the most powerful mechanisms for driving down the cost of environmental improvement projects. The lower the cost, the more such projects will get done. So, remember, the longer the term, the lower the annual payments on loans!

9 Credit Enhancement

If you work for a water/wastewater system with a collection rate of more than 99% and an AAA credit rating, you might be tempted to skip this chapter. Don't. You will need it within the next decade. Trust me.

The words "credit enhancement" have a slightly strange ring to them. We all have credit ratings that tell us how good our credit is. Credit ratings are some kind of analog to our history of debt payments. So, if they are a form of history, how does one "enhance" history?

The words "credit enhancement" are indeed strange when you try to apply them to people. They really don't apply to people. However, they have a definite application to institutions and institutional debt.

From time to time, governments, businesses, and institutions have trouble paying their debts. When they default or are delinquent with debt payments, their credit ratings suffer. Sometimes, the problems are transient and can be readily solved. However, sometimes, with both institutions and governments, the financial problems are systemic and deadly serious. In the late 1970s, the City of New York went bankrupt. Recently, the City of Detroit did the same.

So, what is "credit enhancement?" Essentially, "credit enhancement" is a euphemism for reducing the risk of nonpayment on debt. In general, credit enhancement is used on (nonfederal) government debt and on some major corporate and institutional debts.

If investors believe there is a high risk of loss, they will demand high interest rates to compensate for the risk. They will also only offer short terms to minimize the time in which something can go wrong. Reducing the risk of default on public debt results in longer terms and lower interest rates. Longer terms and lower rates mean lower annual payments (APs). Lower APs mean that more projects get done, and the ones that are done have greater chances of success.

Take the case of a Bus Rapid Transport (BRT) project in a major international city. (The BRT projects are climate change projects, since they can significantly reduce the pollution caused by automobile traffic.) The money to repay debt issued to finance the project can only come from two sources: fares and subsidies from the city. If the annual debt service payments (ADSPs) are high, then the fares must be high or the subsidies must be high, or both. High fares mean lower ridership, which means shooting yourself in the foot, since getting people out of automobiles and on to BRTs is how you reduce green house gas emissions and retard climate change. That is the point of the whole BRT project. By the same token, higher BRT subsidies means that some other vital area of government, such as public health, education, and housing, is getting less money. High interest rates and short terms spell a lose–lose situation for climate change projects like BRTs.

So, the purpose of credit enhancement is to reduce the risk of default or delinquency on debt. Moreover, the goal of using credit enhancement is to drive down the ADSP, or cost, on project debt.

There are nine major strategies for enhancing credit or driving down the cost—or ADSP—of project debt. All of these strategies involve removing—or at least making very remote—the risk of financial loss. As we have said, short terms and high interest rates can doom many environmental infrastructure projects. However, it should be noted that although all these strategies are widely used, some of them are used much more internationally than in the United States. In addition, many of the strategies such as "second loss reserves" are used in financing infrastructure projects with much higher risk profiles—such as urban mass transit or toll roads. That said, it should also be noted that in the future—with the new game in town, as you will read in Section III of this book—water and wastewater systems will be asked to finance projects that are not directly supported by rate bases.

STRATEGY #1—LOCK BOXES

"Lock Boxes" are the easiest form of credit enhancement to employ. The phrase "lock box" is just a shorter way of saying "revenue segregation." If you are a water/wastewater system that sends out its own bills and directly receives payments from rate payers, consider this: have a national bank to do your collections. Don't collect the money yourself.

Furthermore, have the bank collect the money in a special account. Finally, instruct the bank to hold sufficient funds in that account to make any and all ADSPs due that year.

What you are doing here is assuring your creditors/bondholders that the first dollars you collect will go to them. The bank will not transfer funds into your operating account (or any other account) until "all" of the money is there to pay your debts that year.

Bondholder/creditors like lock boxes because they minimize the possibility that there won't be enough funds to pay then. They also minimize the possibility that any of the money will be diverted to other uses instead of paying them.

STRATEGY #2—LIENS

The reason that water/wastewater system debt is so highly regarded is because ratepayers religiously pay their water/sewer bills. When you need to make a capital improvement to your treatment plant, and you borrow money and add the ADSP into your rates, your ratepayers faithfully continue to make their payments. Enough said.

Now, take the case where your local government is concerned about defective septic systems. Or—as part of their stormwater strategy—they want homeowners to replace the asphalt on their driveways with porous, or permeable, pavement. Or, they want homes along waterways to be raised up on stilts (pillars) to protect them against extreme weather events. In addition, the local government wants you to administer these programs. Now what? Is someone who is forced to replace his septic system going to repay the debt for this with the same religious zeal that normal ratepayers pay their bills? The point is: who knows?

The answer to this dilemma is liens. Put a lien on the property. Then, if the homeowner doesn't pay his debt to you, you can legally move against his property. This is especially important with small commercial borrowers. The strip shopping center that replaces the asphalt in its parking lot with permeable pavement just might not pay you next year. In that case, you can move against the shopping center itself to recover your missing funds. This is much safer than an unsecured loan to the owner.

This strategy will require the cooperation of your local government or even your state government. In some cases, a state statute or a local ordinance will be needed to authorize these kinds of liens. However, this concept has a silver lining.

Take the case of a homeowner who wants to spend $15,000 to replace his driveway with permeable pavement. Under normal circumstances, he would go to his bank and get a second mortgage loan. In such case, his bank might give him a 7-year loan at 7% interest. The homeowner's monthly payment on this second mortgage loan would be $226.39.

Now, the pavement will last for more than 30 years. So, your program could offer the homeowner a 30-year loan at your cost of funds, which today should be about 3.5%. In such case, the homeowner's monthly payment would be $67.96. So, think about it. How many more people would install permeable pavement if it only cost $68 a month versus $226 a month? That's the point of programs like these: the lower the cost, the more projects get done and the greater the environmental benefit.

STRATEGY #3—TRANCHING

This is one of the basic principles of credit enhancement. It is essentially the principle of insurance. It begins with a concept that involves organizing financings into "tranches." This is a French word "tranche," meaning slice. The concept here is to slice a financing into pieces with ascending orders of risk of loss. This may sound complicated, but it really isn't. Let me explain.

Let us say there are 1000 apartment units in various suitable places. Now, the probability that "all" 1000 tenants will make full and timely rental payments in any given month is close to zero. Of the 1000 tenants, at least a few will be sick, dead, broke, forgetful, or gone. Conversely, however, the probability that all 1000 tenants will default on their payment is also close to zero.

What is the probability that one tenant will not make his payment? Very high. Close to 100%. What is the probability that five tenants will miss their payments? Again, very high. Again, close to 100%.

What is the probability that 100 tenants—that is, 10% of the total—will miss their monthly rental payment. Very low. Very low, indeed.

Now, let us say that you want to buy these apartments—all of them. The current owner will sell them—all of them—for $70,000,000. Let us say that the monthly rent for each apartment is $600. Let us also say that maintenance, taxes, and insurance are $1,200,000 a year or $100 per month per apartment.

Now, here's a new wrinkle. You are going to have all of the tenants pay their rent directly to a "lock box" at a trustee bank. You will instruct the trustee bank to take $100 out of every tenant's monthly payment and put it in an operating account for taxes, maintenance, and insurance. You then instruct the trustee bank to put the balance, $500 per tenant per month, into an escrow account. This means that if every tenant paid his rent in full, you would have $500,000 a month in rental income in your escrow account. That is a total of $6,000,000 a year.

Now, you want to finance the purchase of the buildings. So, you go to the bank and tell them that you can pay $500,000 a month. You'd like a 20-year mortgage and you've heard that the going interest rate is 6%. So, you ask for a $70,000,000 mortgage.*

What do you think the banker will say?

I think the banker would ask you a politely rhetorical question, something like: "And, how much of a down payment are you planning on making?" Of course, if the buildings cost $70,000,000 and you asked for a $70,000,000 mortgage, you obviously didn't plan on making "any" down payment. If the banker had agreed, you would have scored an infinite leverage deal. You would have bought the buildings entirely with OPM—"other peoples' money." This doesn't happen… at least with banks.

Bankers like down payments. They like to know that their borrowers have their own skin in the game. One of the major reasons for this is, as we know, the possibility that all $500,000 will materialize, that is, the possibility that all 1000 tenants will pay promptly each month—is zero. The banker knows this too. So, the banker is not going to give you $70,000,000.

Now, let us say that you take the same deal to a smart investment banker. The investment banker will know, as we said above, that the probability that 100 tenants, or 10%, won't pay their rent is very, very small. So, he issues a new instruction to the trustee bank: put "the first 900 tenant payments into a new account called" Escrow A. Since the probability that at least 900 of 1000 tenants will pay is virtually 100%, the investment banker knows that he can get a good interest rate on these bonds. He can get a 5% interest rate. In addition, in the bond market, he knows that he can get a 25-year term.

So, instead of one bond, he has two; or more correctly, he has one bond with two tranches. There is the "A Tranche" of 90% of the purchase price, or $63,000,000. The ADSP on the A Tranche is $4,470,000.

Now, what about the other $7,000,000?

Well, now, the investment banker creates another tranche, this time called the "Z Tranche."† The Z Tranche is $7,000,000. However, because it carries virtually all the risk of loss, it requires a shorter term and a much higher interest rate.

So, the Z Tranche will have a 20-year term and will carry an interest rate of 15%.‡ It will also be a level principal payment bond.

Remember that the trustee bank is putting $6,000,000 a year into escrow accounts, of which $4,500,000 is going into Escrow A account. Please note that after paying the ADSP on the A Tranche, there will be $30,000 left in this account, which is your profit.

Next, the investment banker will also instruct the trustee bank to put the other $1,500,000 of rental income into a second new escrow account, the Escrow Z account.

Now, the first year's ADSP on Tranche Z includes 15% interest on $7,000,000, or $1,050,000 plus the level annual principal payment of 1/20th of $7,000,000, or

---

* A $70,000,000 debt for 20 years at a rate of 6% requires a monthly payment of $500,000.

† By convention, investment bankers refer to the tranche that carries the lion's share of the risk of loss, the "Z Tranche."

‡ The Z Tranche is an "equity" tranche (see Chapter 14).

about $350,000 for a total payment of $1,400,000. So, your profit in year one on this $70,000,000 transaction is $30,000 from the A Tranche and $100,000 from the Z Tranche. However, your return on investment (ROI) is infinite, since you didn't invest a penny of your own money! You bought $70,000,000 worth of buildings entirely with OPM!

However, from day one, it gets even better...

Below is what the cash flow looks like on the Z Tranche for the first 10 years, assuming, of course, that the A Tranche has already been paid and you had already put $30,000 per year in your pocket:

Year	**1**	**2**	**3**	**4**	**5**	**6**	**7**	**8**	**9**	**10**
Rents	1500	1500	1500	1500	1500	1500	1500	1500	1500	1500
Interest	1050	998	945	892	840	788	735	683	630	578
Principal	350	350	350	350	350	350	350	350	350	350
Payments	1400	1348	1320	1230	1140	1050	960	870	780	690
Profit	100	152	205	258	310	362	415	467	520	572

What you have just witnessed is a form of credit enhancement! The only problem with this technique is that the Z Tranche is very expensive.

Let's look at another example closer to our subject.

Instead of 1000 rental apartment units, think of 100 climate change projects, each costing $100 million, for a total of $10 billion, and originating in 100 different countries, none of which is in default on any of its foreign loans. Now, investors might believe that at least 90 of these loans are completely secure: absent a global economic meltdown, they will never default. However, they might feel that between 0 and 10 of these projects might get in trouble. So, one might be tempted to break these projects into two tranches, with the A Tranche funded by inexpensive, long-term debt and the Z Tranche funded by costly equity.

Below is an example of the cost of debt and equity with a debt-equity mix of 90:10. The debt is for a 20-year term at a 5% rate. The ROI for the equity is 25%.

	Equity Tranche (25%) ($)	**Debt Tranche (5%) ($)**	**Total ($)**
Investment (Total $1000)	100	900	1000
AP required	25	72.22	97.22

You can tell intuitively that something is wrong here. The equity eats up almost 25% of the ADSP but provides only 10% of the needed cash, whereas the debt furnishes 90% of the cash for only about 75% of the ADSP.

Remembering that our goal in environmental finance is to bring the greatest benefit to the largest number of people at the lowest possible cost, we need to find a way to get rid of the equity component. We must replace the equity.

Enter a concept we will call a "Self-Funded Reserve" (SFR).

STRATEGY #4—SELF-FUNDED RESERVES

Municipal revenue bonds in the United States are often structured with SFRs. Here, the borrower puts up a reserve against its own default. In the United States, this is done singly. This means that if a water system issues a bond, it generally adds to the principal one full year's debt service and places it into the hands of a bank trustee. This is to protect against transient problems. A major water main break is a good example. Here, the problem might be so severe and so acute that it eats up all of the utility's cash reserves and the money it had put aside for its next ADSP. In such case, when the trustee bank does not receive the system's payment on the due date, it dips into the reserve fund that had been set up. Theoretically, once the system gets its repairs done and puts its fiscal house back in order, it should replace the reserve funds that had been spent.

On the international scene, SFRs are very different. The mechanism I just described for the United States works for one project and one bond. However, on the international stage, it would work at the fund level, that is, a fund composed of a hundred or more projects.

Here, each project would contribute 10% to a reserve fund; however, in this case, it would be a "common" reserve fund. In the example of United States, the reserve protected against the borrower's own delinquency. In this new climate change example, "each borrower's reserve fund protects against the defaults of each and every project in the fund." So, in our example, each of the 100 projects would contribute to a common (self-funded) reserve. Please note that "common SFRs" are largely unnecessary in financing traditional water/wastewater projects. However, they will certainly come into play as systems are called upon to finance the more exotic types of water projects.

An SFR is, essentially, an over-borrowing that is placed into an escrow account to pay for defaults. Thus, if a project's cost is $1000, the borrower borrows $1100 and puts the additional $100 into an escrow account. The SFR is fully amortized by the APs of the borrower. This raises the cost to the borrower but not nearly so much as does equity. (Internationally, private equity plays a much larger role in public finance than it does in the United States. Private equity is very, very expensive.) If a borrower borrowed $1100 and put the $100 into an escrow account—instead of using 10% equity—his AP would increase to $88.27* but far less than the combined $97.22 cost of the $900 debt and $100 equity.

Below is a comparison of ADSPs of various levels of equity and the corresponding costs of an equivalent SFR.

The first column is the percentage of debt in the financing. The second column is the complementary amount of funding coming either from equity or from an SFR. The third column is what the AP would be if equity were used instead of an SFR. The fourth column is what the AP would be if an SFR were used instead of equity. The fifth column shows what the difference would be between a financing with equity and an identical funding with an SFR. The percentage difference is also noted.

* Using the same 20-year term and 5% interest rate as in the above example.

Debt (%)	Equity/SFR (%)	AP with Equity ($)	AP with SFR ($)	Difference
90	10	97.22	88.27	$10.03 (10.1%)
80	20	114.19	96.29	$18.07 (18.6%)
70	30	131.17	104.32	$23.31 (25.7%)

It is clear that an SFR can replace equity in a project financing at a much lower cost.

There are two additional benefits of SFRs. They provide debt investors additional protection against loss "over time." They are invested in interest-bearing accounts, and they grow. Furthermore, as each loan repayment is made by the borrower, the outstanding balance on the debt declines. The combined effect of a growing reserve and a shrinking debt balance is a significant reduction in the investors' risk of loss over time.

Here is what the first 5 years of this phenomenon look like, compounding the reserve balance at a rate of 3% per year.

Year	1	2	3	4	5
Reserve balance ($)	1067	1032	995	957	916
Bond balance ($)	103	107	109	113	116
Coverage ratio (%)	9.66	10.28	10.98	11.77	12.65

In the above example, after 10 years, the balance on the bond would be $682, whereas the reserve would have grown to $134, offering almost 20% coverage. After 15 years, the bond balance would be only $382, whereas the SFR would have grown to $156, providing over 40% coverage.

There are some political benefits in the SFR concept as well. If the 100 developing countries borrow an extra $1 billion on top of the $10 billion that they need for their projects and they default, then the money that they lose is their own. There is a certain fundamental principal of justice here: if you ask for help, you must first be willing to help yourself. All investors feel more comfortable if a project sponsor (borrower) has its own skin in the game. The SFR is just that: the borrowers' own skin in the game.

So, Strategy #4 is: replace equity with SFRs, wherever possible.

STRATEGY #5—SECOND LOSS RESERVE

Strategy #5 is much like an SFR, except that it is used to buffer losses once a primary—or first-loss—fund/guaranty is exhausted. Very little probable use for this is in the US water/wastewater market. In the international market, however, an international development bank, such as the African Development Bank, might take a 5%–10% first-loss position on a large fund financing many projects. However, the credits might be so shaky that the bankers put together a second-loss reserve fund to further protect investors. Again, this has not much probable use in the US water/wastewater market.

STRATEGY #6—TAX REVENUE INTERCEPTS

Think unrelated tax receipts!

These are credit support mechanisms that can be used to assuage investors' fears of nonpayment. It is important to note that these mechanisms do not replace usual income streams. They are not the source of repayments. Rather, these mechanisms are standbys, fallbacks, or safety nets, just in case revenues are off or the subsidies aren't enough. They serve as guaranties. Here are some examples.

In the late 1970s, the City of New York was bankrupt. Nonetheless, it needed to borrow money to undertake needed civic infrastructure projects. However, no one would lend to the city. At that time, there was a sales tax in New York State that was shared with local governments. The tax was 5% in those days, 3% of which went to the state and 2% of which went to the City. The City collected all the taxes incurred in the City and forwarded the State's share to the State Treasurer.

To overcome the City's credit problem, the City and State enacted laws to provide that all sales tax revenues in the City would be paid into an account in the City's name at an international bank acting as a trustee (a "lock box"!). So, the City, itself, never legally touched this money. The bank forwarded the State's share to the State Treasurer. However, the bank held the City's share for the benefit of its lenders. (As the City dutifully made each payment on its debt, the bank trustee would release the funds it then held into the City's general account.) So, the City was then able to borrow money by pledging the hundreds of millions of dollars of its sales tax revenues—in the hands of an independent bank fiduciary—to its new lenders. So, the City's local sales tax revenues were "intercepted" for the benefit of its creditors.

As it turned out, the City never had to use any of these "intercepted" sales tax revenues to repay its debt. It was able to make all payments out of its other general revenues. The "intercept" was truly a standby, fallback mechanism.

This intercept mechanism is a powerful device that can be used to reduce the risk of nonpayment to lenders.

India has an even more powerful variation of the local tax intercept. This mechanism has been successfully used to support local water system debt in the State of Tamil Nadu. Here, certain tax revenues are "owned by the local governments," but they are "collected by the state governments." In this case, the local government and the state government entered into an agreement with the local government's lenders such that if the local government ever failed to make a timely loan repayment, the lender could go to the state, which would "intercept" enough of the local government's tax revenues to make up the missed payment to the lender. A similar tax-revenue-sharing structure exists in Turkey.

In Mexico, certain tax revenues are legally shared by the federal and state governments. Because both the federal and state governments participate in the benefits of these monies, they are called "participaciones." These "participaciones" are collected by the federal government. If a state borrows money, it, of course, promises to repay the money with interest from its general revenues. However, in addition, the state may pledge that in the event that it fails to pay in the normal course of business, the lender may apply to the federal government, which is legally obligated to "intercept" the state's "participacion" of the tax and pay the state's lenders whatever they are owed.

Standard & Poor's (S&P) reported that in 2009, 80% of all municipal bonds issued in Mexico were supported by "participacion" agreements.

There is another variation of tax intercept that might be called a "super tax intercept." ("Super" in the sense of "above.") Recall in the example of the City of New York, above, that the State of New York also owned 3% of the 5% sales tax revenue. This money belonged to the State, not the City. Nonetheless, if the State really wanted to assist the City, and the State did not want to issue its own guaranty (or was itself not creditworthy enough to issue a guaranty as in the case of Argentina), then the State could have agreed that its share of the sales tax revenue could have been pledged to the repayment of the City's debt.

As noted above, these credit support mechanisms are not the source of debt repayment; rather, they are a standby or a fallback in case the borrower cannot make a debt payment. However, they are exceedingly powerful. They ease investors' fears of nonpayment to the point where they will accept lower interest rates and significantly longer terms on their investments. This results in lower project debt payments from local governments and breaks the upward spiral of risk and cost.

Tax revenue intercepts (TRIs) are widely used abroad. However, they started in the United States in the 1980s with school district debt, where a given state had enacted a student per-capita aid formula. In such case, if a school district defaulted on its bond issue, the bondholders could go to the state treasurer, who would be legally required to "intercept" the per-capita aid payment to the school district and pay it to the bondholders (as much as they were owed) instead. The TRIs can provide powerful support to debt, helping to reduce interest rates and lengthen terms.

STRATEGY #7—EXTERNALLY FUNDED GUARANTIES

Strategy #6 involves guaranties that are funded with cash; however, unlike the case with SFRs, the cash comes from another source, generally a government program that funded.

At more than $110 billion, the US Clean Water State Revolving Fund (CWSRF) is the largest dedicated environmental finance program in the world. It is just such an externally funded source of guaranties.

All 50 states and the Commonwealth of Puerto Rico participate in the CWSRF program, which are managed at the state level. Since its inception in 1987, Congress has funded the CWSRF at various levels. These funds are appropriated to the US Environmental Protection Agency (EPA) and then distributed to the states under a congressionally approved formula. Under the Clean Water Act, in order to receive CWSRF funds, each state must appropriate $1 for every $5 it receives from EPA.

Today, the "size" of the CWSRF is more than $110 billion, which means that more than $110 billion of projects have been funded. The "net assets" of the CWSRF are more than $40 billion. "Net assets" equals: (a) the congressionally contributed funding, plus (b) the 1 : 5 matching funds appropriated by the states, and (c) interest earned on funds that have been loaned. The difference between the program "size" (more than $100 billion) and its "net assets" is called "leverage." The CWSRF's leverage ratio is only about 2 : 1.

Standard & Poor's has studied the default histories of municipal bonds for almost a century. The default history on bonds issued to fund public wastewater projects is approximately 0.04, or 2500 : 1. In January of 2011, S&P published a document, requesting comment on, "inter alia," a new total leverage ratio for anyone guaranteeing public wastewater project debt. Standard & Poor's said that if the guarantor wanted to maintain its highest investment-grade credit rating (AAA), the maximum leverage ratio could be no higher than 75 : 1.

With more than $40 billion of net assets, at a 75 : 1 leverage ratio, the CWSRF has the potential to leverage $3 trillion.

Why is the CWSRF important to a discussion of credit enhancement? This is specifically because the 51 CWSRFs have the legal authority to issue guaranties. Theoretically, their collective guaranties, up to a total of $3 trillion, would be rated AAA.

Now, what is most important to realize is that the guaranties offered by the CWSRFs are not government guaranties in the legal sense. No, when a CWSRF issues a guaranty, the underlying documents specifically provide that the only source of the guaranty is the CWSRF's net assets. In other words, these CWSRF guaranties are not the legal promise of any government; rather, they are guaranties backed by enormous amounts of cash.

When you think of "externally funded guaranties" like these, think of a bank Letter of Credit (LoC). What is the bank's LoC backed with? The bank's equity; its own funds. Here in our example, the CWSRF guaranties are backed by each CWSRF's net assets. No legal promises; just cash.

STRATEGY #8—LEGAL GUARANTIES

A legal guaranty is an enforceable contract, whereby one party promises to make full payment for the debt of a second party if the second party fails to make a full and timely payment on such debt. If you lend $1000 to your neighbor and the neighbor's rich uncle guaranties the debt in writing, and if your neighbor fails to make an agreed-upon payment at an agreed-upon time, you can take your written guaranty from the rich uncle into court and compel him to pay up.

Another matter to consider about guaranties is the financial relationship between the guarantor and the debtor.

If I walked into an investment banker's office and told him that I wanted to issue a bond for $10 million, he would undoubtedly be both incredulous and dismissive. However, if I handed him a letter from the US Secretary of the Treasury, saying that the US government would guaranty my $10 million bond, he would be very impressed and get right to work. I would wind up with a long term and a very low interest rate on my bond because of the guaranty.

Now, let's reverse the situation. Let us say that I agreed to guaranty all of the US Treasury's outstanding debt. That's about $17 trillion dollars. Do you think that would result in lower interest rates for treasury bonds? I think, not.

So, the point here is that the value of any guarantor is directly related to the strength of his or her credit.

Legal guaranties are a sore subject with most governments. The reason is very straightforward. When a government, say, the City of New York, guaranties a debt, the amount of that debt is almost always counted 100% against the credit of the guarantor, that is, the City.

When we say, "counted against the credit," we mean two things. First, if there is a constitutional debt limit, the entire amount is counted against the guarantying government's legal debt limit. So, when a government guaranties a debt, it is burning up its available borrowing power. It might also borrow the money itself and relend it.

Second, guarantying a debt counts against the credit of the guarantor from the perspective of the international credit rating agencies as well. This has two ramifications.

First, when credit agencies look at governments (or corporations), they assess their overall financial capability. In the case of a city, they look at the wealth of the community. They look at the tax rates. They look at the local government's budget. Moreover, they come to a conclusion as to how much that governmental unit can borrow with a certainty of full and timely repayment. If the government's outstanding debt is below that number, and the government and the community are otherwise in good order, then the government will be awarded an AAA rating. The less this is true, the lower will be the credit rating. Once a credit rating gets below investment grade, which is BBB-, the guaranties become pretty useless.

This brings us to our second point about legal guaranties and credit ratings that many of governments—all over the world—that have serious need for many types of capital projects have execrable credit ratings.

So, what is important to remember is that the value of the legal guaranty is based on the credit strength of the guarantor.

STRATEGY #9—MUNICIPAL BOND INSURANCE

Strategy #8 is to use municipal bond insurance to enhance your credit, if you can find it and if the math works out in your favor.

Before the subprime mortgage crisis, there were 15 municipal bond insurance companies, which, in 2005, guarantied almost 60% of the about $400 billion of municipal bonds issued that year. In 2011, there was one company, and now, there are four.

This entire industry was born in 1971, with the creation of the American Municipal Bond Assurance Company (AMBAC). This company insured only municipal bonds! The second entrant in the field was the Municipal Bond Insurance Association (MBIA) in 1973, which was founded by a consortium of five major insurance companies: Aetna, Fireman's Fund, Travelers, Cigna, and Continental. The MBIA also insured only municipal bonds.

This industry, composed of only two companies, churned along making considerable profits until the mid 1980s, when the first signs of trouble began to appear: competition.

From the mid 1980s to the 1990s, this little industry swelled from 2 companies to 15. Word got out about how much profit AMBAC and MBIA were making, and other investors flocked to join the field.

With 15 companies now in the market, the second sign of trouble appeared: there were now 15 companies competing for business. The natural thing to do, especially for the new guys trying to get a toehold in the business, was to cut premium rates. As premium rate cutting got worse, profits plummeted and investors became disillusioned.

Now, the third sign of trouble appeared: collateralized debt obligations (CDOs).

Frantic over falling profits, one by one, the companies were lured away from insuring traditional municipal bonds, where there was too much competition fighting for too few deals, into the heady realm of CDOs. These CDOs housed subprime mortgage obligations. There was no money to be made in insuring traditional municipal bonds, but there were plenty of profits to be made in insuring CDOs. So, the old municipal bond insurance industry, which now called itself the financial guaranty insurance industry, lunged headlong into the uncharted waters of insuring CDOs.

To say that they made a mistake is one of the great understatements of the century. The financial guaranty insurance companies totally miscalculated the risks inherent in these CDO with their payloads of subprime mortgages. However, the industry was not alone. Even the rating agencies miscalculated the risks and continued to affirm the AA and AAA ratings of these financial guaranty insurance companies, even as their risk portfolios became more and more bloated by the subprime risks. Finally, the entire empire collapsed. As the 100+% loan-to-value subprime mortgages came up for rate adjustments, people began to realize that there were thousands of homeowners who could not afford the new, higher rates. When they couldn't pay their mortgage that had been sold into a CDO by the originating bank, the CDOs started defaulting on "their" payments. When this happened, investors went to the financial guaranty insurers. The financial guaranty insurance companies had nowhere near-enough reserves to pay such massive losses. So, the financial guaranty insurance industry collapsed.

However, like the phoenix that rises from its own ashes, the industry will be back. There was one company in 2011. Three more have opened their doors in last few years.

So, should you use municipal bond insurance to enhance the credit of your transaction? If it works for you, yes. What does this mean? This means that if the bond insurance will lower your interest rate more than enough to cover the cost of the insurance premium, then you should use it.

Because of the checkered history of bond insurers, their insured bonds sometimes do not trade as well as an uninsured bond with the same rating. In other words, if City A, with an AA credit rating, issues a bond, and City B, with only a BBB rating but that buys insurance from an AA-rated insurer, issues an identical bond, you will often see City A's bond trading at a better interest rate than City B's bond trading.

So, the appropriate way to deal with bond insurance is to determine at what rate the two bond insurers bonds are trading. Then determine where your bonds would trade without insurance. Then request a quote from the two insurers and determine if the lower payments due to the higher credit rating will more than offset the cost of the insurance.

In general, the rule is that the insurance premium should eat up just under half of the spread between what your bonds would trade at with the insurance versus without the insurance. In other words, if your bonds would sell at 5% without the insurance and 4.5% with the insurance, then the premium should cost you about 0.2%, which means your all-in rate with the insurance would be 4.7%. So, you would save the difference between 5% and 4.7%, or 0.3%.

So, “credit enhancement” isn’t just some slick Wall Street terminology. Rather, it is a term that covers methods and mechanisms to lower your interest rate (and possibly lengthen your term) by improving the credit rating on your project.

Whenever you are involved in financing projects, your goal will be to drive down their cost. As such, you should consider any of these credit enhancement strategies that will help.

First and foremost, as has been expressed several times in this book, avoid using equity, if at all possible. It is so expensive.

Second, use lock boxes and liens. Third, divide your financing into tranches. Fourth, look for external guaranty or debt insurance funds. Fifth, look for a credit-worthy government that is willing to guaranty the financing. Sixth, check our whether financial guaranty insurance is available and whether it is cost-effective to buy it.

Always remember that our overarching goal is to provide the greatest environmental benefit to the largest number of people at the lowest possible cost. Credit enhancement is a means of achieving that goal.

Section II

The Basics of Financing Traditional Water and Wastewater Projects

10 The Clean Water State Revolving Fund

Chapter 1 of this book is entitled "The 800-Pound Gorilla!" It discusses the $3.7 trillion municipal bond market, and you understand that the entire municipal bond market is used to finance schools, roads, and almost any other type of government facility. In other words, as you saw, water and wastewater projects make up only about 15% of the total market.

That said, how would you describe a $3 trillion fund that was just limited to wastewater projects? You'd have to call it a ONE-TON GORILLA!

Is it possible that the Clean Water State Revolving Funds actually have $3 trillion of capacity? Yes, theoretically, it is possible, but in reality, it is at least $1 billion. How so?

Standard & Poor's (S&P) has studied the default histories of municipal bonds for almost a century. The default history on publicly owned treatment works (POTWs) is approximately 0.04, or 2500 : 1. On January 24, 2011, S&P published "Request for Comment: Bond Insurance Criteria." This document sets forth proposed revisions to the criteria under which it rates financial guaranty insurers, especially municipal bond insurers. Paragraphs 29–31 of that document describe a new test for total maximum leverage for various credit ratings—which is the most radically conservative test ever suggested by an international credit rating agency. Table 9 under Paragraph 31 states: "Maximum Leverage Consistent with a 'AAA' Rating (for) Risk Category—US municipal: 75 : 1." This means that a collective fund like the Clean Water State Revolving Fund (CWSRF) could guaranty 75× its net assets with an AAA rating. So, at a 75 : 1 leverage ratio, with net assets of $40 billion, the CWSRF could guaranty some $3 trillion, that is, $3,000,000,000,000, of bonds to pay for water pollution control.

As you will note, the data S&P studied were classic POTW bond issues—bonds that are financed by ratepayer payments. However, SRFs finance more than just this. They finance nonpoint source projects and estuary projects that aren't necessarily supported by ratepayer payments. So, perhaps a leverage ratio of 75 : 1, yielding $3+ trillion of financial capacity, is a bit too rich of an estimate. Let's drop it all the way down to 25 : 1. Even this means that the combined CWSRFs have total available leverage of $1 trillion.

So, should we rename this chapter a gorilla of, maybe, 500 pounds? Actually, no. Because the unfortunate matter is that CWSRF leverage is wildly underutilized. Instead of a robust $1 trillion program, it is only about 1/5th that size or around $200 billion. That's too bad for America's waters.

As I said at the beginning of this book, the Clean Water Act was enacted over the veto of President Richard Nixon in 1972. Embedded in it was a "construction grant" program that pumped more than $70 billion into sewage treatment plants. Moreover, this was back in the day when $70 billion was real money! Matching funds from the states brought the total up to close to $100 billion.

Then, as often happens with grants, EPA began to notice that some projects were overbuilt. "Golden faucet" stories started popping up. There were fears that because the money was free, some of it was being wasted. Then entered Ronald Reagan, a man who never liked grants or "free money" anyway. So, in 1987, the gods came together and enacted a wholesale amendment to the Act, replacing the construction grant program with the Clean Water State Revolving Fund. Now, EPA was still making grants, but they were making them on two major conditions. First, that the states come up with $1 for every $5 they got in grants from the federal government, and, second, and most importantly, that the states "lend" the money to their wastewater utilities and local governments. Loans! No more grants! And, so it began, the CWSRF!

Today, some 29 years later, the 51 CWSRFs have provided more than $110 billion of financial assistance to about 36,000 projects. It is by far and away the single most successful environmental finance program in our galaxy.

POINT SOURCE POLLUTION

There are three main authorities for making loans under the CWSRF. The first are traditional loans to sewage treatment plants under Section 212. Some 96% of CWSRF money has gone to finance sewage treatment plant projects. Water in the United States is now visibly and noticeably cleaner that it was in 1972, thanks mostly to the almost $200 billion we've put into sewage treatment plants in the last 44 years, more than half of which came through the CWSRF.

The CWSRFs work in annual cycles. Each year, all states are required to create Intended Use Plans (IUPs) to tell the world and EPA how they are going to spend their money. Point-source projects must be specified and listed. Nonpoint sources may be listed in categories.

Once the EPA regional office approves a state's IUP, the state CWSRF then begins the solicitation process for actual projects in detail. These projects are scored on the basis of cost and effectiveness and then prioritized. They are then published on a Project Priority List (PPL). The trick here is that if the state has $200 million worth of projects on its PPL and only $100 million to spend, then the "first" $100 million of projects—"the $100 million of highest projects on the PPL"—get funded. These projects are usually called "above the line." The projects that aren't going to be funded—unless an "above the line" project drops out—are called "below the line." In many states, the competition is fierce to get projects "above the line."

Sewage treatment works are point sources of water pollution. So are the effluent pipes that come out of factories and businesses, many of which provide thousands of jobs. All point sources of water pollution are issued National Pollution Discharge Elimination System (NPDES) permits by their respective states under authority of the Clean Water Act. The NPDES permits are how the states control point-source pollution. Only POTWs are eligible for CWSRF funding. The business and industrial polluters don't get a cent from the CWSRF. There are some limited exceptions, which you will see below in certain estuaries. However, on one of these days, some politician is going to go ballistic, because a large industry—with several thousand jobs—in his district is closing its doors because of the cost of adding more pollution control technology.

The average big business might be able to get a pollution control loan from their bank for 5 years at, maybe, 7%–9% interest. A $10 million bank loan for 5 years at 8% will cost the company $2,505,000 a year. If they were eligible for the CWSRF, they could get a 30-year loan at 3.5%, in which case the annual payment would be $543,000 a year. That almost $2,000,000 savings might mean being able to stay open and save those several thousand jobs. Someday, the politicians might just learn….

NONPOINT SOURCE POLLUTION

The second authority for CWSRF funding is for nonpoint pollution sources under Section 319 of the Act. It should be noted that under the Act, states—through their NPDES permits—can "mandate" that both public and private point source polluters "must" take certain remedial actions. However, it is not so with nonpoint sources of pollution. No NPDES permits. No power to compel polluters to remediate or take any actions at all.

Each state is required to prepare a "319 Program" from time to time that identifies major categories of nonpoint source pollution and to describe strategies to address them. These strategies should then be included in the state CWSRF's IUP.

How do these projects get "above the line?" How do they get funded, especially if there is fierce competition among the point source projects? There is no straight answer to this question. Remember that 96% of all CWSRF funds have gone to POTWs. So, it seems likely that the state boards that vote on their respective CWSRF projects simply add in a few—probably high-profile or politically popular—nonpoint source projects for good measure.

NATIONAL ESTUARIES

Section 320 of the Act creates the "National Estuary Program." The purpose of this section is to get states to organize programs to protect major estuaries. So, first, the Act lists 28 "National Estuaries" by name (see Appendix A). The Act empowers the states to create boards to administer the estuaries. It specifically requires the NEP agencies to create "Comprehensive Conservation and Management Plans" (CCMPs) and to keep them updated. The CCMPs may specify projects that are needed to maintain or improve water quality in the estuary. These projects may even be privately owned point sources. (This is the exception, noted above, to the point source funding exclusion for privately owned facilities.) The CWSRFs are specifically authorized to finance projects identified in their states' CCMPs.

FUNDING METHODS

Direct Loans

Of the 37,000 instances of CWSRF financial assistance in the last 28 years, all but about a dozen of them have been in the form of direct loans. Until 2014, the maximum term for a direct loan was 20 years. Now, it is 30 years.

Interest rates are an interesting matter for discussion under the CWSRF. None of the states charges market rates of interest. Fifty CWSRF programs offer "subsidized" rates of interest. Vermont charges 0% interest.

In most of the states that make subsidized loans, the interest rates are usually about 50% of the AAA rate for long-term municipal bonds. Many states have deeper subsidies for their poorer areas.

Why subsidized interest rates? These revolving funds would "revolve" faster if they charged market rates. The answer—almost unbelievably—lies in the fear of competition. The 800-Pound Gorilla, known as the municipal bond market, is the CWSRFs' great competitor. So, what's the problem?

The problem lies in some small additions that were made to the Clean Water Act for reasons totally unrelated to clean water.

The first and most infamous of these is the application of the Davis–Bacon Act, which specifies, in each state, the wage level that skilled craftsmen must be paid under CWSRF-financed projects. These wage levels are often high enough, so that when the labor component of a particular project is large enough, it would make it uneconomic for a community or wastewater utility to borrow from the CWSRF rather than going directly to the municipal bond market, where there is no Davis–Bacon requirement. So, the CWSRFs insist that they must offer subsidies in order to offset the higher costs imposed by the Davis–Bacon requirements.

Now, the Davis–Bacon requirement is called a prevailing wage law. Thirty-two states have their own prevailing wage laws, which may be higher or lower than the Davis–Bacon. For these states, the Davis–Bacon requirement shouldn't present much of a problem. However, for the 18 states that have no prevailing wage laws, the difference in Davis–Bacon wages rates and local wage rates may pose a real problem (see Appendix B for the states with no prevailing wage laws).

Another additional cost of the CWSRF is the requirement that its projects use American Iron and Steel. This might be a problem in certain circumstances, but the Clean Water Act was amended in 2014 to add an exemption from this provision if the cost of the American Iron and Steel was 25% higher than foreign sources.

The final complaint that CWSRF managers have to deal with is compliance with a host of federal "crosscutters"—requirements that apply to any project of any kind that receive federal financial assistance. They are listed in Appendix C. Take a look at them. Don't they suggest a bureaucratic nightmare? No wonder, some systems try to avoid going to the CWSRF for funds. The municipal bond market has none of these requirements.

So, the CWSRF managers may, indeed, be right that they must offer subsidies to make up for the imposed costs and for the bureaucratic complexity of the program.

Loan Guaranties

Of more than 37,000 instances of financial assistance by the 51 CWSRFs, about a dozen have involved loan guaranties. Most importantly, only one—(Yes, one!)—of more than 37,000 projects involved the use of a financial guaranty in the municipal bond market. In August of 2014, the New York State Environmental Facilities Corporation, which manages the CWSRF in New York, guaranteed a $23.4 million municipal bond issued by a sister agency, the New York State Energy Research and Development Authority. The proceeds of the bond went to finance energy-efficiency—(Yes, energy efficiency!)—projects in homes and small businesses. As you will see in Chapter 20, energy efficiency reduces demand for electricity largely produced by power plants that spew vast amounts of nitrogen into the air. This airborne nitrogen winds up in out waterways when it rains. EPA has a center at the University of Illinois that studies air deposition in receiving water bodies.

Purchase of Local Obligations

Rather than making a direct loan, a CWSRF could also buy a bond issued by a local government or wastewater utility issued to finance an eligible project. Before 2014, some 20 states used this authority. Their primary reason for doing so was to obviate the 20-year limit in the direct loan program. In other words, there is no time limitation in this section of the Clean Water Act, so when a project needed to be financed for 30 years, rather than making a direct loan, the state SRF would simply purchase a "local obligation" from the local government or utility. In 2014, however, Congress amended the Clean Water Act and removed the 20-year limitation from the direct loan provision. So, now, the SRFs don't need to use the local obligation purchase section to extend financing terms to 30 years.

Investments

The Clean Water Act gives the CWSRFs the authority to "invest" funds. It doesn't say where or why. The CWSRFs have traditionally interpreted this authority to apply to funds that weren't currently in use. For example, when a CWSRF receives a loan repayment from one of its borrowers, there may be many months, or even a year or so, before those funds are re-loaned to another borrower. So, in the meantime, they can be invested. Most of them are invested in traditional securities such as US Treasury Bills.

A few states, such as Massachusetts, use this investment authority to further their program. Many small systems don't have ready access to funds to prepare projects. For many projects, sophisticated studies must be done and equipment must be designed. All of this costs money, which they may not have. So, some states use their investment authority to furnish this type of "pre-project" funding to their local governments and wastewater utilities. There is another way in which this investment authority could be used—very creatively.

In Part III of this book, you will read about some of the new twenty-first century challenges that CWSRFs are facing. One of them, I call the "upstream-downstream" problem. This is where polluted water is flowing downstream to communities that need it for drinking. These communities have to spend money to purify the water before they can drink it. Sometimes, this purification can be very expensive. Moreover, sometimes, it can be much cheaper to fix the problem upstream. The problem is that, as you know, there is no authority in the Clean Water Act to compel nonpoint polluters to take any remedial action. Furthermore, almost all of such nonpoint sources are agricultural and are privately owned. There are certainly no provisions of the Act to compel farmers to pay for any kind of nonpoint source pollution.

Now, take the case of Maryland and Pennsylvania. The EPA's Chesapeake Bay Program reports that 60% of the pollution in the Bay comes from agricultural runoff. Much of this comes right down the Susquehanna River from the many farms along the river in Pennsylvania. Now, Harrisburg, the capital of the Commonwealth, is the last major city on the Susquehanna, before it flows into Maryland.

Let us say that Harrisburg wants to reduce the upstream pollution. This would greatly benefit Maryland too, which is further downstream.

So, as an inducement for the City of Harrisburg to provide grants to farmers to reduce upstream pollution (at a lower cost than Harrisburg would have to pay to clean the water), could Maryland offer to "invest" in a Harrisburg bond to fund this strategy? Maryland could offer Harrisburg a much lower rate from them to do so. If the going rate for bonds was 4%, Maryland could offer Harrisburg, say, 1%.

Is this possible? The answer is yes. There is no geographic limitation on the CWSRF's investment authority. They can invest out of state.

So, has this ever been done? No, of course not. Will it ever be done? As the upstream–downstream issue becomes more apparent to more managers and more policymakers as the twenty-first century progresses, the answer will probably be yes.

So, the CWSRF is the single most effective water pollution control finance program in the world. However, it has well more than $1 trillion of unused capacity. As the twenty-first century rolls on, more of this capacity will undoubtedly be used.

11 The Drinking Water State Revolving Fund

The Drinking Water State Revolving Fund (DWSRF) was created by Congress in 1996. It was modeled on the highly successful Clean Water State Revolving Fund (CWSRF), which was enacted in 1987.

Through 2014, the federal government has made contributions to the states in the form of capitalization grants of $17.3 billion. As is the case with the CWSRF, the states must match these funds on a 1 : 5 ratio. So, the states have added another $3.46 billion to the pot. All in all, from 1996 to 2014, the DWSRF has provided $27.9 billion financial assistance to more than 11,400 projects.

In almost all respects, except as noted below, the DWSRF operates just like the CWSRF, and in most states, the two programs are staffed by the same office. Like the CWSRF, the states must prepare annual Intended Use Plans (IUPs) to demonstrate how their funds will be allocated. These IUPs are submitted to EPA. Then, again, as is the case with the CWSRF, the states must prioritize the applications they receive and prepare a project prioritization list. The DWSRF also has the capability to make direct loans, guaranty loans, purchase local obligations, provide for insurance for local debt, and invest its funds.

However, unlike its older sibling, the DWSRF can make special provisions of additional subsidies for small, truly disadvantaged communities beyond a 0% interest loan. These additional subsidies can take the form of grants, principal forgiveness, and negative interest loans. So, you see, unlike the CWSRF, the DWSRF funds do not always revolve. There is a limit of 30% in the Safe Drinking Water Act (SDWA) for these "additional subsidies."

Like the CWSRF, the DWSRF's direct loans are limited to 20-year terms, except in the case of a disadvantaged community (as defined), for which these loans can be for up to 30 years.

The DWSRF has six basic eligibilities. These are treatment facilities, transmission and distribution facilities, development or rehabilitation of wells or other sources of water, water storage facilities, the consolidation of two or more drinking water systems, and the creation of new systems.

There is one feature of the DWSRF that is not at all like the CWSRF: multiple set-asides. Both the SDWA and the Clean Water Act (CWA) allow the states to set aside 4% of their capitalization grants to use for program administration. However, the SDWA specifically provides that up to an additional 27% of the capitalization grant that states receive may be "set aside" for three other specific purposes.

Up to 15% of the 27% may be used for four specific purposes. The first is to acquire conservation easements for the protection and preservation of source waters. Second, these funds may be used for other, voluntary (incentive-based) measures for source water protection. Third is to provide technical assistance to a water utility as part of a state capacity development program. A "capacity development program" is essentially a training program to assure that each system has the (1) technical, (2) financial, and (3) managerial capacity to consistently deliver water in accord with SDWA standards. Last is to finance Wellhead Protection Programs.

The other large set-aside is for what might be called water management programs. Up to 10% of the 27% can be used for these programs; however, there is a catch to this provision. In order to set aside money out of the federal capitalization grant for these programs, the state must come up with additional matching funds. In this case, the matching funds are dollar for dollar. These "water management programs" include the development of an Operator Certification Program. This set-aside can also actually pay for the state's capacity development program. It can also be used to manage or provide technical assistance to source water management programs. To administer the state's Public Water Supply Supervision program, which is the core drinking water program in each state.

The third additional set-aside in the SDWA is 2%, out of the 27%, which can be used to provide technical assistance to small systems, which are defined as systems with fewer than 1000 users.

So, the DWSRF, much like its older sibling the CWSRF, is essentially a low-cost loan program for drinking water. Both programs work pretty much the same way.

12 USDA Rural Development Water and Environment Program

Like the two State Revolving Funds (SRFs), the Water and Environment Program (WEP) of the US Department of Agriculture (USDA) is one of the federal government's most successful programs. However, probably because it only affects communities of fewer than 10,000, it is not very well known.

The program started in 1937 at the Farmers Home Administration (FMHA), an agency of the USDA. The FMHA was subsequently re-titled the Rural Development Administration. Then, for some unknown reason, they dropped the word "Administration." So, now, the agency is just called "Rural Development."

People also used to refer to the program as the "water and sewer" program, since that's where most of the money went. However, as the "environment" gained more caché in official Washington, the USDA decided to change the official name to the "Water and Environment Program." Of course, it has always dealt and still deals with solid waste disposal too.

Now, most farmers are very leery and suspicious of EPA. They realize that a lot of pollution comes from the fertilizer, pesticides, and manure on farmland. So, farmers tend to keep a weather eye on EPA, wondering when EPA will try to bring an enforcement hammer down on them.

Their nightmare came true in June of 2105, when EPA redefined the "Waters of the United States," asserting jurisdiction over streams and lakes far upstream, bordering tens of thousands of farms. The American Farm Bureau Federation immediately sued. As of this writing, there are 10 separate cases, where 9 states and 1 Chamber of Commerce have joined the Farm Bureau as plaintiffs, challenging what is called the "Clean Water Rule." The cases are slowly wending their way through the federal court system, and as of now, the Sixth Circuit Court has stayed the enforcement of the rule.

If farmers can avoid dealing with EPA, even when they need to undertake environmental projects, they do. They'd much rather deal with their friends at the USDA. Notwithstanding the enmity between the farming community and EPA, both the WEP and the two SRFs do terrific jobs in bringing clean water to the American people.

The WEP is both a grant and a loan program. Rural Development touts the WEP as a "needs-based" program, and it truly is a needs-based program. It provides assistance to only those communities and utility systems that cannot obtain credit from commercial lenders or investors.

The general formula for distributing loan and grant funds among the states is based on three criteria: (1) the state's percentage of the national rural population, (2) the state's percentage of the national rural population with incomes below the poverty level, and (3) the state's percentage of national nonmetropolitan unemployment.

Despite having this complicated, clearly needs-based formula, the spending patterns can be very unusual. For example, in 2013, California got $29.3 million of loan funds and $7.7 million of grant funds. However, in 2014, California got only $10.6 million of loan funds and $2.2 million of grant funds.

In 2014, South Carolina received the most loan funds, some $55 million, along with $13.2 million of grant funds. That same year, New York, Michigan, and Kentucky, each received more than $45 million of loan funds.

The WEP uses three strategies to deal with its "needs" basis. The first involves the interest rates they charge. The program has three different interest rates that they charge, depending on market conditions. At today's writing, the AAA interest rate for 30-year tax-exempt municipal bonds is about 3.2%. The WEP's "Normal Rate" today is 3.125%. Their "Intermediate Rate" is 2.5% and "Poverty Rate" is 1.825%.

The second strategy involves the term of their loans. Assuming that the assets that they are financing have at least 40-year service lives, the WEP will make 40-year loans. The annual payment on a 30-year loan of $1 million at the Poverty Rate (1.825%) is $43,583. The payment on the same loan with a 40-year term is $35.443. As you can see, the annual payment on the 30-year loan is 23% higher. So, the 40-year term is vital to the WEP's goal of delivering low-cost water services to rural communities.

The third strategy involves the use of their grants to buy down the cost of a project.

The 47 state WEP offices plus the national staff in Washington closely scrutinize the fiscal impact of their loans on the people in the rural communities that have to pay the bills. The WEP staff look at the water/sewer rates that their clients are currently paying, as well as what they would have to pay to finance the project at the three rates of interest that the WEP offers. The staff then look not only on the direct impact but also at the water/sewer rates that other ratepayers are paying in adjacent districts. When they find that their project will have a seriously adverse financial effect on the ratepayers of the project district, or that those ratepayers will be paying significantly more than the ratepayers in neighboring districts, they then use their program grant funds to buy down the cost of the project—to make it more affordable for the ratepayers.

Here's how all this works. A $1 million project at the Normal Rate (3.125%) with a 30-year term would require an annual payment of $51,847. At the Intermediate Rate (2.5%), it would cost $47,778 and at the Poverty Rate (1.825%), it would cost $43,583.

By extending the term to 40 years, the following annual payments would be required by using the 3 different WEP interest rates: $44,141; $39,836; and $35,443, respectively.

Now, here is what the annual payment would be if the project size was bought down with a 30% grant ($300,000) *and* the term was 40 years *and* the Poverty Rate of interest was charged as $24,810.

As you can see, by employing these 3 strategies to their fullest extent (which is rare), the WEP can reduce the cost of projects by more than 50%—from $51,847 down to $24,810. That is why the WEP is such a powerful program to bring clean water and safe drinking water to rural America.

The WEP was funded at a $1.2 billion level in both 2013 and 2104. Of these amounts, about 70% was for loans and 30% was for grants. It should be noted that unlike the CWSRFs and DWSRFs, when communities make their loan payments for the WEP, the money goes back into the US Treasury. As you know, when repayments are made to the SRFs, the money stays in the state fund and is re-loaned to other borrowers. The $1.2 billion of assistance that the WEP provided in 2014 went to about 1,000 borrowers. So, the average WEP project size was about $1.2 million.

The WEP also has loan guaranty authority, as do the two SRFs. However, of the $1.2 billion that the WEP received in 2014, only some $7 million was used for guaranties. So, much like the SRFs, the WEP's real power lies in its loan authority and, of course, in its grant authority.

The combined grant and loan authorities of the WEP with a funding level in the $1.2 billion range make it one of the most powerful and effective programs in rural America.

13 Other Water and Wastewater Finance Programs

This chapter will briefly describe the finance programs offered by the National Bank for Cooperatives and the National Rural Water Association, as well as four programs offered through the state rural water associations in Pennsylvania, Kentucky, Kansas, and Minnesota.

CoBANK

The National Bank for Cooperatives (CoBank) has provided more than $1.5 billion of loans as financial assistance to water and waste disposal systems. Their borrowers include not-for-profit associations, municipalities, and investor-owned water utility companies. CoBank offers pre-development loans, interim or bridge loans, and long-term loans for capital projects. They also re-finance existing debt and finance leases. CoBank is part of the Farm Credit System.

NATIONAL RURAL WATER LOAN FUND

This National Rural Water Loan Fund was founded by the National Rural Water Association, with a grant from the US Department of Agriculture (USDA). Local governments, including water and wastewater utility districts, Native American tribes, cooperatives and not-for-profit corporations, are eligible. This program finances pre-development costs on an interim basis. The program can also finance equipment replacement, system upgrades, or small-scale extensions of service. Emergency loans for disaster recovery are available for up to 90 days with no interest. Energy-efficiency projects can also be financed to lower system costs.

The application process is straightforward, and the turnaround time on loan approvals is very quick, usually only a matter of a few days.

Loans are limited to the lesser of 75% of the project cost or $100,000, whichever is less. Terms can extend to the shorter of the service life of assets being financed or 10 years. Interest rates are set by the USDA through the Rural Development Water and Environment Program.

KANSAS

The Kansas Rural Water Finance Authority (KRWFA) was created under the auspices of the Kansas Rural Water Association. The KRWFA provides financial assistance in the forms of interim loans, construction and equipment financing, and the financing and re-financing of loans and bonds. The KRWFA provides this financing to (1) rural water districts, (2) public wholesale water supply districts, and (3) cities. The KRWFA has issued more than $100 million of bonds.

PENNSYLVANIA

The Pennsylvania Rural Water Association (PRWA) has assembled an array of financial services for its members. These include loans, leases, and bonds.

Of special note is PRWA's bond program, which is actually a bond pool program. PRWA notes that issuing a tax-exempt municipal bond for less than $5 million is not cost-effective because of high legal and underwriting costs. So, PRWA organizes bond pools for members with projects below $5 million, where major costs can be shared. These pooled bonds can have terms of up to 30 years, assuring the lowest annual payments. The PRWA also has a working capital loan program and a construction loan program. So, presumably, systems can take out either a construction loan or a working capital loan to pay for their project and then keep those loans outstanding until there are enough projects from other systems, so that they can all be funded together through a pooled bond.

In addition to working capital loans and construction loans, the PRWA also offers straight project loans with terms of up to 25 years. It also offers equipment loans and debt refinancing.

In terms of leases, the PRWA offers equipment leasing. It cites the following benefits for such leases: no referendum requirement, tax-exempt rates, not carried on the books as debt, quick approvals, no expensive legal opinions, flexible terms, and financing for full service life of equipment.

The PRWA offers a truly impressive array of financial services for its members.

MINNESOTA

Minnesota also offers an impressive array of finance programs for its members. In 1999, the Minnesota Rural Water Association founded the Minnesota Rural Water Finance Authority. The authority operates an interim loan program for systems that have received commitments from the USDA's Rural Development for a loan through its Water Environment Program. The authority funds these interim loans by selling 12-month tax-exempt notes. If the borrower's project is not complete within this period, it is rolled over into the next note. The interest rate is flat to the borrowers, but if the program makes a profit, borrowers get rebates on the interest they have paid.

Minnesota also has what it calls a "Midi" Loan Program, which is a "quick and low-cost alternative to conventional general obligation (GO) bond sales." This provides tax-exempt financing for very small communities for projects of less than $1 million. Midi loans are for communities with populations of 400 or less, where the per capita debt—both outstanding and planned—is less than $5,000. Midi loans require GO pledges from the borrowing communities.

Minnesota also offers a "Micro" Loan Financing Program for small projects of between $30,000 and $250,000 that can be repaid in 7 years or less. These loans are structured as a straight GO note and are tax-exempt. Micro loans with terms of 2 years or less can be paid off at anytime at par. If the term is more than 2 years, it can still be paid off within the first 2 years; however, the borrowers are charged a 0.5% call premium. These micro loans are described as "simple, quick, and low-cost."

Finally, Minnesota also offers what it calls its "Mega" Loan Financing Program, which is for borrowings of more than $1 million. These are traditional tax-exempt municipal bonds but with expedited and simplified procedures.

KENTUCKY

In 1995, the Kentucky Rural Water Association founded the Kentucky Rural Water Finance Corporation (KRWFC) to provide two specific types of financing for its members. The KRWFC provides interim construction loans for systems that have received commitment letters from the USDA's Rural Development for a loan through its Water and Environment Program. This is structured as a loan pool with flexible terms that is financed with tax-exempt notes. There is a stated rate of interest, but borrowers get rebates when the KRWFC makes a profit on the pool. As of 2016, the effective borrowing rates were in the 2%–2.5% range. Since the program began, it has provided more than $838 million in construction loans to 385 members of the association.

The KRWFC's other major program is its Flexible Loan Program. This program offers loans with terms of up to 35 years. It requires a GO pledge from its borrowers and is funded as a pooled tax-exempt bond issue.

The program has some very attractive provisions. Approvals can be issued in only 2 days. Funds can be made available in 35 days. The program offers both fixed and variable rates. Its bonds are rated AA–, which assures very low interest rates. The program uses one investment bank and one bond counsel, which greatly simplify issuance procedures and reduces costs. Cost of issuance is estimated at only 3%. Moreover, the program permits the re-financing of existing debt.

Since 1995, the KRWFC has made loans to members of the association for 211 projects, totaling more than $339 million.

***

As you can see, the NRWA's program as well as the four state programs are largely an effort to supplement loans offered by the Water and Environment Program of the USDA's Rural Development. They provide simple, flexible, and inexpensive procedures that are well suited to the needs of small rural water systems. In short, they provide small rural water and wastewater systems with the same type of access to the $3.7 trillion American municipal bond market that only the largest systems have traditionally enjoyed. As such, they provide rural America with great benefit at low cost.

Section III

The New Game in Town

14 Public–Private Partnerships

Public–Private Partnerships (P3s) have become very popular—at least to talk about.

There seems to be a lot of confusion about P3s. People speak of them as if they were some kind of magic silver bullet to solve public finance problems. They're not. So, we need to start off by getting a few things straight.

First, a P3 isn't a finance mechanism; it is simply a legal structure; in this case, it is one for implementing infrastructure projects. Second, P3s aren't new. They've been around as long as government and the private sector have co-existed. Third, they aren't necessarily innovative or creative. When a village hires a local trucking firm to pick up the garbage, this is a P3. However, it is hardly brilliant, creative, or innovative.

On the other hand, there are some brilliant and creative P3s. Take certain toll roads, for example—not the usual toll roads, but some of the new ones where there are two lanes of traffic going in the same direction. One is a toll lane; the other is free. When traffic is slack and the free lane is virtually empty, the toll lane costs little or nothing. However, at rush hour, when the free lane starts to choke up, the toll lanes start getting very expensive. The denser the traffic in the free lane, the higher the charge in the toll lane. This works very well. It works well when the income stream supporting the P3 is very elastic. After all, you may not want to pay \$10–\$15 to get home at a reasonable hour, but unless you really love sitting in traffic, you're going to do it.

However, this isn't the case with most infrastructure projects, especially water and wastewater projects. There, the income streams are anything but elastic. The days of pay toilets have come and gone. Think of water rates, sewer rates, and stormwater fees or charges—all are very inelastic. Water and wastewater systems don't reset their rates every few minutes, or days, or even months.

The recent twist on P3s is this relatively new association with finance. Some people speak of P3s as if they can produce money out of nowhere for infrastructure projects. The hidden assumption here is that the "out-of-nowhere" is actually some secret sources of money in the private sector. This is a very dangerous idea.

Government has a moral mandate to provide infrastructure at the lowest possible cost to the people. So, the finance question with P3s is: which of the Ps is bringing the money to the table for the project. Is it the public P or the private P? This question is crucial.

If the public P brings the money, then the financing will generally be done through the tax-exempt municipal bond market. This enormous \$3.7 trillion market assures that the lowest possible interest rates and the longest possible terms will be available. This means the lowest possible cost for the people.

For example, a $10 million wastewater project in today's world might result in a $10 million municipal bond with a 30-year term and a triple-A interest rate of 3.5%. This would cost the county taxpayers $544,000 per year—a very low cost for such a project.

However, if the private P brings the money to the party, then the story can be entirely different.

Whether a P3 meets the moral mandate to produce the lowest possible cost depends on the private P's appetite for return on investment (ROI) and term or exit strategy, that is, how fast they want their money back.

If the private P is a charity or other donor, then, there is no problem. The ROI is 0%, and the project will, indeed, carry the lowest possible cost. However, very few private Ps are charities.

If the private P wants an ROI north of 15% and wants out after 5 years, the project will be very, very expensive. As of May 19, 2015, the US Treasury reported that the most recent yield on its 30-year bonds was 3.02%. So, at a time when treasuries are a hair over 3%, what crooks, what criminals, what villains, and what charlatans would demand an ROI of more than 15%? Some Wall Street pirates, no doubt!

How about the California Public Employees Retirement System? Yes, the government employees in California!

The California Public Employees Retirement System, better known as CalPERS, is the largest pension fund in the United States. It has $294 billion of assets. A total of $31.3 billion, or a little over 10%, is invested in "private equity." CalPERS reported that its benchmark return rate for its private equity investments for the fiscal year ending June 30, 2014, was 15.4%. What's more, private equity investors usually want out after 5–7 years.

So, if CalPERS had financed our $10 million wastewater treatment project at 15.4% for 5 years, it would have cost the county taxpayers $3,011,000 a year. If you were on the board of directors of the wastewater utility and you had to vote on either CalPERS money or municipal bond money, which would you choose: $3,011,000 per year or $544,000 per year?

What about CalPERS "other" money? What about the 90% of their assets that aren't tied up in "private equity" investments? Sure, CalPERS will accept less than 15% on its other investments. It all depends on the risk involved, as well as on the term.

So, let's say there was a modest income stream supporting a project that could be done with a P3—almost like a municipal bond, but not a municipal obligation.

In this case, CalPERS or any other retirement fund, sovereign wealth fund, or life insurance company might accept, say, a 6% rate of interest. However, they wouldn't want their money tied up for more than 10 years. Most of the nongovernment debt market is 10 years or less. So, let us consider a $10 million P3 financing at 6% for 10 years. This would result in an annual cost to the system's ratepayers of $1,359,000 per year. Again, if you were a board member, what would you vote for: $1,359,000 or $544,000?

So, P3s definitely have their uses. They are great for garbage collection. They work well when the income stream supporting them is very elastic, as in toll roads or some airports. However, the majority of water and wastewater infrastructure projects are supported by fixed payment streams. In this case, beware of P3s. They are no silver bullets.

In Chapter 19—Resiliency Projects, you will see further discussion of some specific uses of P3s. You will also see some specific analyses of the cost consequences of certain P3s, depending on the appetite of the private "P" bringing the money to the table for their required or desired ROI.

15 Nonpoint Source Projects

Forty-four years ago, when the Clean Water Act was first passed, municipal sewage was the number one source of water pollution. Today, stormwater and agricultural runoff are the two major culprits, and, of the two, agricultural runoff is the worst (See Chapter 16).

As you know, the Clean Water State Revolving Fund (CWSRF) has provided over $110 billion of financial assistance in its history of more than 25 years. As you also know, 96% of these funds have gone to wastewater treatment plants. This means that about $4.5 billion has gone to "other" projects. Most of these "other" projects involve nonpoint source pollution.

Now, if you are a sewage treatment plant or a private company that discharges effluent, you are a "point source" of pollution and will have to have a National Pollution Discharge Elimination System (NPDES) permit. These permits are issued by states. The authority to do so is found in the Clean Water Act. This is how the states control the water quality within their borders. They can legally compel permit holders to reduce the pollutants they discharge.

Clustered around cities are municipal and industrial point sources of water pollution regulated by NPDES permits. Urban point source pollution control projects are readily financed. Industrial polluters can finance through banks or through the corporate bond market. Public sewer systems can issue municipal bonds or borrow from the CWSRF. Private bonds and loans are readily repaid from corporate sales revenues. Public systems repay debt through their massive ratepayer bases.

However, if you are a nonpoint source of pollution, there are no permits. Furthermore, there is no authority in the Clean Water Act for states—or anyone else—to compel you to do anything. The state cannot compel you to reduce pollution from your nonpoint source.

So, reducing nonpoint source pollution is voluntary—sort of. Some of them are voluntary, for sure.

Of the $4.5 billion "other" projects that the CWSRF has financed, some $1.48 billion, or about one-third of this money, has gone to cap or line sanitary landfills. These are expensive projects. The $1.48 billion paid for 385 projects, or a little over $3.8 million for each project. Capping or lining sanitary landfills is not voluntary at all. They are required by local laws and ordinances that govern the disposal of solid waste. Now, solid waste is itself a major source of pollution—but not water pollution. However, when it's piled together in open landfills and then it rains, it becomes a source of water pollution. Despite the fact that the source of the pollution is one "point"—the landfill—it is classified as a nonpoint source. It may be regulated by state or local laws or regulations, but it is not regulated by the Clean Water Act.

About $2.3 billion of the $4.5 billion has gone for a variety of projects such as silviculture, urban projects, groundwater protection, mining cleanups, brownfield cleanups, leaking underground storage tank problems, hydromodification (water channel changes), and decentralized sewage treatment. A little over $100 million has gone for unclassified projects.

The remainder, which totals only $705 million, or 16% of the "other" money, has gone for agricultural runoff. What is truly astonishing about this statistic is that $705 million were paid for some 8939 agricultural runoff projects, which means that the average cost of a single project was $78,868. What is even more astonishing is that virtually 100% of these projects were voluntary. This means that 8939 farmers borrowed an average of $78,868 each for their water pollution control project, and, although they may have received additional subsidies in the form of some principal forgiveness, they agreed to pay it back out of their own pockets!

As remarkable as "voluntary" is, it's no longer enough.

If we are going to get serious about reducing water pollution from agricultural runoff, we are going to have to design financing programs that don't rely on volunteers. We can't expect farmers to pay for all of this out of their own pockets. Farms are an absolute necessity. They are not a luxury or a hobby. To the extent that they create water pollution in the process of creating food for the rest of us, we all benefit from the food; therefore, we all need to share the cost of reducing the pollution.

This is the challenge of the twenty-first century in the water quality business.

There are two major types of agricultural runoff: pollution from croplands and pollution from livestock.

Of the 8939 agricultural runoff projects financed by the CWSRF, 3364 dealt with "Agricultural Animals" (read "manure") for a total of $254 million, or $75,505 per project. The other 5575 projects were "Cropland" projects and totaled $451 million, or $81,000 per project.

If the CWSRF has provided over $110 billion to over 36,000 projects, you can readily see that the average project size is about $3 million. Moreover, the dominant borrowers are publicly owned treatment works (POTWs), all of which have professional staff whom the CWSRF staff can work with. In other words, the CWSRFs are neither equipped nor staffed to deal with thousands of farmers who need $75,000–$80,000 apiece.

So, across the country, two dominant forms of nonpoint source lending have developed. Both are employ intermediaries that not only have financial roles but also serve as aggregators. (You will see in Chapter 18, which deals with energy-efficiency projects—that are even smaller in size—that aggregators/intermediaries are desperately needed.)

In the agricultural world, both local banks and local governments such as counties and conservation districts serve as both intermediaries and aggregators.

If the intermediary is a bank, the financing will be done through a "loans-to-lenders" or "linked deposit" program. Here, the farmers go to banks that are part of the linked deposit program and apply for the loans they need. The bank approves and takes it to the CWSRF. The SRF makes a deposit in, or buys a certificate of deposit (CD) from, the bank, which is equal to the loan amount. Furthermore, the CWSRF agrees to accept a lower-than-market interest rate on its deposit or CD, which the bank must pass along to the farmer. If the farmer defaults on his loan, the bank is on the hook to the SRF. In this way, the SRF gets the whole issue of dealing with hundreds or thousands of small loans off its plate. This is a popular way to finance nonpoint source projects among the state CWSRFs. There is an excellent example of a nonpoint source project—actually a watershed preservation project—financed by a linked deposit program in Annapolis, Maryland, in Chapter 17.

The other way to finance nonpoint source projects through the CWSRF is for the conservation district to do all of the loan application work with the farmer and then go to the SRF. However, then, instead of the SRFs making the loan directly to the farmer, they make the loan to the conservation district, which then on-lends it to the farmer. So, the conservation district takes the responsibility of disbursement, collection, and loan monitoring. Technically, if the farmer defaults on his loan from the conservation district, the district is still on the hook to the SRF. However, this has never happened. Or, if it has happened, no one has reported it.

The Delaware SRF has an innovative take on this concept. It conducts its "AgNPS" program through the state's conservation districts, but it also requires that participating dairy and poultry farmers have a professional affiliation with major producers in their respective industries. On the dairy side, farmers must be affiliated with Dairy Farmers of America, Inc., Land O'Lakes, or the Maryland and Virginia Milk Producers Association, Inc. On the poultry side, they must be affiliated with Allen Harim LLC, Amick, Inc., Mountaire Farms of Delmarva, Inc., or Perdue Farms, Inc. These affiliations have nothing to do with regulation; they are part of a marketing strategy. Delaware uses these agricultural associations and affiliations to get the word out about their animal waste loan program. Very smart idea!

So, here is the situation as we approach the second decade of the twenty-first century.

Upstream on all the farms, you have the #1 cause of water pollution: agricultural runoff. In addition, downstream, you have thirsty people clustered in cities, who need clean water. They don't need the polluted water that is coming down the stream toward them.

Upstream are the nonpoint sources. In comparison with a municipal sewage treatment plant, they contribute small amounts of pollution. However, a city or town may have one or two sewage treatment plants, whereas there may literally be thousands of nonpoint sources of pollution upstream. It is these nonpoint sources of pollution that are the challenge of the twenty-first century. There are so many of them. They are almost all on private land, and most importantly, this private land is almost entirely owned by families or private individuals. How is one family supposed to pay for a water pollution control project? How are farmers supposed to pay for pollution abatement projects on their land?

This is the "upstream/downstream" problem. Moreover, it is the water pollution control finance challenge of the twenty-first century.

Financial strategies and mechanisms need to be devised to address these questions in a manner that is fair and equitable—and, most importantly, acceptable to all. Issues such as paying for constructed wetlands and bioreactors—devices that reduce agricultural runoff—need to be examined, especially in the context of nutrient trading and offset regimes.

Here is how one small community in Montana solved its upstream/downstream problem.

The forested lands of Haskill Basin nestled in the mountains of northwest Montana are home to about 75% of the drinking water of the City of Whitefish. At the top of the basin is the Big Mountain Ski Resort.

Whitefish, Montana, is a community of about 8700 souls in about 3300 households at the bottom of Haskill Basin. The city used to get its water from three creeks, First Creek, Second Creek, and Third Creek, which are all tributaries of Haskill Creek, which drains Haskill Basin. By the 1970s, spray irrigation from the resort's wastewater lagoons, plus septic system problems at private ski chalets, plus parking lot runoff, had caused the city to close its water intake in First Creek.

In 2015, there were about 3020 acres of undeveloped woodlands located in Haskill Basin and owned by a private lumber company. As the lumber company sat with its massive, prime acreage, the pressure to sell it for development grew every day and the people of Whitefish realized this. Development of this acreage would impair all of Whitefish's water supply.

The city and the lumber company reached an agreement for the city to buy the 3020 acres for $20.6 million. There were about $700,000 of associated expenses for a total project cost of $21.3 million. Of this amount, the company agreed to donate land valued at $3.9 million as a tax write-off. The city was also able to obtain both a $7 million Forest Legacy Grant from the US Forest Service and a $2 million Cooperative Endangered Species Conservation Fund grant from the US Fish and Wildlife Service. The city also applied to the Montana CWSRF, managed by the Montana Department of Environmental Quality, for an $8.4 million loan with a 10-year term and a 2.5% interest rate.

Now, the annual payment on such a loan would be about $960,000, which would be a burden—to say the least—on the 8700 residents of Whitefish. As it turns out, Whitefish was able to come up with another source of revenue to repay the loan.

Montana law provides that counties may enact resort taxes of up to 3% in their respective jurisdictions. Whitefish had enacted a 2% resort tax. At a 2015 referendum, the voters approved of—with 84% of the vote—raising the tax to 3% and pledging the proceeds to pay for the CWSRF loan.

The city has estimated in its FY'16 budget that the proceeds of the increase in the resort tax will be sufficient to pay the annual debt service on the CWSRF loan. However, just in case, the city has also pledged its water rate payments as a secondary source of funding.

So, of the total $21.3 million needed for the project, $8.4 million came from the CWSRF loan, $9 million came from grants from the US government, and $3.9 million was donated by the lumber company that owned the property.

In Chapter 8, you learned the importance of "term." So, it should be noted that the annual payments on a 30-year loan of the same amount and at the same rate that Whiteface got from the SRF would be only $401,000, instead of $960,000. The city could easily have applied for its loan with a 30-year term, instead of the 10-year term. The Department of Environmental Quality has subsequently indicated that they would have approved a 30-year request if it had been made.

So, as you can see, the solution to Whitefish's upstream/downstream was solved with a large, dedicated—if external—revenue stream in the form of the sales tax increase on resorts. So, it was largely the people who came to ski from out-of-state who paid for the project. The people wrestling with coastal resiliency projects, as described in Chapter 19, should take heed of this idea to get out-of-staters to share the burden!

Let me give you an example of how an upstream/downstream problem can be dealt with. Let us say that there is a soybean farmer in Iowa who wants to improve his crop yield by installing tiles along the ditches that drain his fields. The tiles improve the drainage, and the improved drainage means more crops. For 100 acres, this might cost $500,000. Unfortunately, tiling drainage ditches creates more runoff. So, the farmer's project will have a negative impact on the water quality of the receiving stream or lake. The farmer would have to finance his tile project through his local bank. Now, considering the farmer a good, reliable customer, the bank might offer him a 10-year loan at 5% interest. This means that the farmer will pay $64,752 a year for his tile project.

Now, let's say that the clean water folks from the Department of Environmental Protection have a word with the farmer. They suggest that he set aside 2 of his 100 acres and build a wetland to absorb the extra nutrients that his tile project will create. As a matter of fact, a 2-acre constructed wetland will absorb far more than the increased nutrients caused by the tile. So, there would be a net increase in water quality.

Now, the cost of the constructed wetland will add $50,000 to the project, so the total project would then be $550,000 instead of $500,000. BUT—and this, indeed, is a big "but"—now, since there is a net increase in water quality, the entire project may be eligible for financing from the CWSRF. This means that the farmer could finance the entire $550,000 project for at least 20 years at a (nonsubsidized) rate of about 3.5%. In such case, the farmer's annual payment would be $38,700. This is a lot better than $64,752. Moreover, if the CWSRF determine that the service life of the project will be 30 years, then the annual payment would only be $29,904.

Now, notice above how we said, "the entire project *may* (emphasis added) be eligible." Well, the rule of thumb with the CWSRF has always been that if there are two components of a project—one eligible and the other ineligible—then the SRF can finance the eligible component but not the ineligible component. However, in our example, if the SRF doesn't finance the ineligible "component," "there won't be an eligible component." Why would any farmer in his right mind pay the bank $64.752 a year and then go to the SRF to borrow more money for the "eligible" component, which means he would have to pay back even more money? No, if the SRF doesn't finance both "components," then the eligible component just won't get done. What will be the effect if the SRF refuses to fund the whole project? Well, then, only the ineligible component will get done, which will result in a net reduction in the state's water quality. So, a very strong case can, and should, be made for having the SRF finance the entire project.

As you can see, it is these types of new strategies that need to be adopted to improve the quality of upstream water before it becomes a downstream issue.

There is an especially important strategy that was pioneered by the State of Ohio's CWSRF and has now been adopted by Iowa, Oregon, and Idaho as well. It is called a "Sponsorship Program." Basically, what it involves is an ingenious interest rate adjustment that induces a downstream community to undertake an upstream project, not at its own expense, but at no expense at all!

Here is a great example of a sponsorship project that was organized by the Iowa CWSRF for the upstream benefit of the people of Sioux City.

In 2013, Sioux City, Iowa, applied to the CWSRF for 3 projects, totaling $14.4 million, that dealt with modernizing and upgrading its traditional sewage treatment system. At the same time, the city was seeing problems in a place called Ravine Park. The problem dealt with the eroding banks of the Ravine Park waterway that polluted the water. Ravine Park was "characterized by steep, eroded gullies." What was needed was a project to shore up these gullies and stream banks to prevent the erosion, which would reduce the pollution coming downstream.

A study was done, and it was determined that the problem could be fixed at a cost of about $1.4 million. But who was going to pay for it?

So, the CWSRF organized the Ravine Park into a sponsored project—sponsored, that is, by the City of Sioux City.

Here's how they did it.

The CWSRF intended to give Sioux City a subsidized interest rate of 2% on a 20-year loan for their $14.4 million. Sioux City's annual payment on this loan would have been $880,657.

Instead, the CWSRF asked if Sioux City would "sponsor" the Ravine Park project, bringing their loan total to $15.8 million. Here's what the CWSRF did. It offered to reduce the interest rate on the Sioux City's loan from 2% to 1.03%. The annual payment on a $15.8 million loan at 1.03% interest for 20 years is $878,209. In other words, if Sioux City would agree to increase the amount of their loan from $14.4 million to $15.8 million—and "sponsor" the Ravine Park project—they would be rewarded with a lower interest rate, which would make their annual payment smaller than it would have been on the original $14.4 million loan!

Isn't that totally ingenious! The upstream Ravine Park project got done ostensibly for free!

Would the Ravine Park sponsorship project work if it were, say, $3 million? The answer is no.

When would this type of sponsorship program work? Here is some math.

If the market interest rate were 4%, then most state SRFs would make their subsidized interest rate loans in the 2% range. Using a $10,000,000 project as the "main" project, how much of a sponsored project the SRF affords to finance? Well, the annual payment on the "main" project itself for 20 years at 2% would be $611,567. If we added a $1 million "sponsored" project, then it would be an $11 million loan. So, let's drop the interest rate to 1% and see what happens. The annual payment on a 20-year loan of $11 million at 1% would be $609,568. So, it works when the "sponsored" project is 10% of the size of the main project.

However, what if the sponsored project were $2 million, or 20% of the size of the "main" project? Then, the total loan would have to be $12 million. Would the sponsorship work at this level? No, clearly the SRF would have to reduce the interest rate even further. To what level would the SRF have to reduce the interest rate to make it work? Well, the answer is that the interest rate would have to be about 0.1%, which is one-tenth of 1%. To be exact, the annual payment on a $12 million loan for 20 years at 0.1% interest is $606,319. So, it works!

In other words, sponsorship projects work when the "sponsored" project is 20%, or less, of the "main" project. However, they do work and can be an invaluable tool for paying for "upstream" projects.

At present, as noted above, only Ohio, Iowa, Oregon, and Idaho offer sponsorship programs. However, as you can see, sponsorship programs can be especially valuable for projects such as watershed protection, where the only project cost that is incurred is in buying forested land. So, the US Endowment for Forestry and Communities has begun a campaign to persuade the other 47 state CWSRFs to adopt sponsorship programs as well.

Other innovative approaches for paying for upstream projects are "nutrient banking" and "nutrient trading" programs.

These programs have a distinctly technical side. They involve quantifying nutrients that pollute water into "credits." Fortunately, this task is undertaken by the states with their environmental scientists.

There are the usual two sides to this equation: those who can create credits and those who either need them, outright (like a developer, as you will see), or are willing to pay for them because they just want the benefit.

Think of a farmer who will plant a few rows of nitrogen-absorbing trees along the banks of the stream that runs across his property. The farmer actively works his field. No matter how careful he is dosing his fields with fertilizer, there is always some that runs off when it rains and winds up in the stream, which ultimately winds up in someone's drinking water intake downstream. The trees, which the farmer plants, absorb the nitrogen, preventing it from going downstream.

The trees are going to cost the farmer $100,000 to plant. He is not thrilled about paying this out of his own pocket. Planting the trees isn't going to improve his crop yield or put more money in his pocket. So, he consults with representatives of his conservation district, who tell him about the nutrient credits he could earn. They suggest that he invites to his farm the scientists from the State's Environmental Protection Administration to determine how many credits he could earn. The conservation district people also tell him about low-cost loans available through a program that the district manages on behalf of the state's CWSRF. So, the farmer investigates the situation.

The scientists come out, look over the proposed project, look at the locations where the trees will be planted, and conclude that the farmer could earn 10,000 nutrient credits for undertaking the project to plant the trees. The farmer also learns that he can get a $100,000 loan from the CWSRF through the program where they partner with the conservation district. The loan would be at a subsidized rate of 2% and would have a 20-year term. The farmer does the math and sees that he would owe the CWSRF $6116 per year in annual payments.

In the course of working with the CWSRF, the farmer learns that the state has organized a nutrient credit trading program. On further investigation, he learns that there is a developer downstream who wants to build a subdivision. However, the sewage treatment plant in the area he wants to build is maxed out on its NPDES permit.

The POTW can increase its capacity only if it gets credits from somewhere. So, the sewer authority tells the developer to go out and buy 10,000 nutrient credits and then they'll hook up his new subdivision to their POTW. He has no choice.

Through the auspices of the CWSRF, the farmer puts his 10,000 credits on the market for $0.70 apiece. That would give him $7,000 per year, and his annual payment to the CWSRF is only $6,116. So, the farmer would actually make a small profit—if he could sell the credits for 70 cents apiece.

The developer does his own math and determines that he will pay up to $6,000 a year, or $0.60 per credit. The developer finds out about the farmer's 10,000 credits being offered at 70 cents each through the CWSRF.

Eventually, again, with the assistance of the CWSRF, the farmer and the developer agree on a price of $0.65 per credit. So, again, the farmer makes a small profit and the developer gets the permit approvals he needs from the state so that he can get his project done.

Now, the above is a very simplified example of a nutrient trading market.

The State of Pennsylvania through its infrastructure finance agency, PennVest, which manages both SRFs, and the Department of Environmental Protection actually operate a nutrient credit trading program. It works by auction.

So, if our farmer and developer were in Pennsylvania, our farmer would have put in his 10,000 credits with a strike price of $0.62. A "strike price" is the point at which the credits are withdrawn from the market. The farmer would do this to assure that he would receive at least $6,200 a year, which he needs to make his $6,116 annual payment to the CWSRF for his $100,000 loan. So, if our trade were in Pennsylvania, then the developer would have bid up to the 65-cent level, at which time, presumably, the other bidders would have dropped out. So, he would have won the bid.

PennVest usually conducts auctions every few months. In both 2014 and 2015, it had four auctions. The results of the last auction in 2015 are interesting to note for three reasons.

In the first of two auctions conducted that day, a town authority and a private corporation each bid and won 20,301 credits that had been offered by Lycoming County. So, in this case, you have both a private entity and a public entity that needed credits, presumably to satisfy their permits. So, you can have multiple parties on either side in the auction.

Second, you have a local government, Lycoming County, that is selling its credits. Let us say that the county bought a large farm that it wanted to turn into a county park. So, it was taking a working farm out of production, which means that it would be eliminating the amount of nitrogen that ran off the farm when it was active. For doing so, the county was entitled to earn nutrient credits. Of course, it was free to sell those credits as well, and so it did.

The third matter of interest was that in the second auction that day, there was a private company that was offering to sell credits. In this case, it wasn't just a company that was undertaking some unrelated activity and just happened to wind up with some nutrient credits in its pocket. No, a major part this company's business was to generate and sell nutrient credits. This company actually buys credits from farmers and resells them through PennVest's nutrient trading program.

This leads us into our final topic, which is very much akin to nutrient banking, but it is called mitigation banking.

Mitigation banking goes back to 1983, when the US Fish and Wildlife Service published the first guidance on the matter. The US Environmental Protection Agency's "Mitigation Banking Factsheet" defines a mitigation bank as "a wetland, stream, or other aquatic resource area that has been restored, established, enhanced, or (in certain circumstances) preserved for the purpose of providing compensation for unavoidable impacts to aquatic resources permitted under Section 404 or a similar state or local wetland regulation." (The "Section 404" referred to here is Section 404 of the Clean Water Act.)

Back in the day when the Fish and Wildlife Service first published its guidance, it was concerned mostly with the impact of road and transportation construction projects that would tear up wetlands—most of the time "unavoidably." Environmental policy makers couldn't stop these construction projects, so they came up with "mitigation banking" as a way to get back at least as many acres of wetlands as they were losing. In other words, if a state highway project were going to tear up 100 acres of wetlands, then someone from the state had to go out and "restore, establish, enhance, or preserve" an equivalent 100 acres of wetlands. Doing so is called creating "ecological offsets." The cost creating the ecological offsets became part of the cost of the highway project.

However, as time went on, other needs for mitigation banking arose.

In our example above about the developer wanting to build a new subdivision, let us say that in order to do so, he needs to destroy 10 acres of wetlands. There is no way he is going to get the permits from the state, unless he "mitigates" the impact of his actions.

Now, originally, nutrient banking was done on an acre-for-acre basis. However, as time went on, the mitigation business became more sophisticated. Instead of acres, the system began to use "credits," much like nutrient credits. The credits were estimated by state scientists and approved by the Interagency Review Team, which is the name of the official group that oversees the entire mitigation process.

So, in the case of our developer, the state scientists will study the developer's site and tell him how many mitigation credits he will need. He will then go to a mitigation bank and buy these credits.

So, what are mitigation banks? These are a relatively new phenomenon. They are basically private firms that have amassed vast amounts of expertise in the restoration, establishment, enhancement, and preservation of wetlands.

Resource Environmental Solutions (RESs) is a private company that is a mitigation banker. It was founded less than 10 years ago in 2007. In 2014, it became the largest provider of ecological offsets in the United States. The company now works in nine states. As you might imagine, a company with this type of expertise works comfortably on both sides of the aisle. It works for both permittees and the regulators who approve or issue the permits. RESs works for land owners as well as not-for-profits and nongovernmental organizations interested in protecting the environment.

The RES(es) of the world are clearly our future. First, think of how valuable such concepts as "credits" and "offsets" are. They enable progress to go on without harm to the environment. Think of the concept of companies like RES with vast competence in providing invaluable environmental services. Think of a regulator looking at a permit application from XYZ Development Inc. that knows zippo about wetlands other than they must be destroyed so that their subdivision can be built. Let us say that the XYZ Developer tells the regulator that it will find an area where it can restore, establish, enhance, or preserve a wetland—and that it will do it by itself. If you were the regulator, would you be happy with this situation? Would you be happy letting a company, like XYZ, undertake an offset project when it'd never done one before? Or, would you be happier if XYZ hired RES to handle the offset project?

The concept of having independent highly technically competent firms as intermediaries between the state and the entity needing to do an environmental project is very attractive and is gaining proponents in many areas.

Energy-efficiency programs (financed through CWSRFs) usually require that the energy-efficiency devices financed be supplied and installed by the contractors whom the state energy office or agency has approved.

The same is true with green infrastructure. It is better that the owner of the shopping center—who knows nothing from horticulture—hire an expert firm to not only install but also maintain its green roofs and the rain gardens scattered across the property.

In addition, as you could see from the players in the PennVest nutrient credit auctions, private, highly qualified firms have roles to play there too.

So, the role of private companies in effecting high-quality environmental improvement projects in the twenty-first century is here now, and it is growing. When it comes to financing nonpoint pollution projects, you had better got used to dealing with companies in the private sector. They are here to stay.

16 Stormwater and Green Infrastructure

In the last 10 years, the importance of stormwater as a source of water pollution has grown exponentially. Stormwater is now a major issue.

The responsibility for dealing with stormwater issues varies from community to community. In some places, it is the responsibility of city or county governments. Their Departments of Public Works handle stormwater. Some other areas, states, or counties have created special stormwater utilities, but in other areas, wastewater utilities have been called upon to deal with this issue.

The goal of stormwater management is to reduce the rush of large volumes of rain into the receiving water bodies that they pollute. Massive flows of stormwater scour the landscape they pass over, picking up all sorts of ugly passengers such as animal waste, chemicals in the ground, and plain old garbage and debris, and send them right into the receiving river, bay, or lake. Before it becomes stormwater, massive rains also scour chemicals such as nitrogen from smoke stacks and exhaust pipes and carry it into lakes and rivers. Reducing these horrific flows is where green infrastructure comes in. Most green infrastructure promotes absorption of rain where it falls. By doing so, it reduces the quantity of rushing stormwater. In addition, by letting rain percolate into the ground, it recharges aquifers and groundwater sources that are critical for uses as drinking water in dry seasons.

There are several types of green infrastructure. They basically involve eliminating or reducing impermeable surfaces such as roads, sidewalks, roofs, and parking lots and replacing them with vegetation, or at least, access to soil that can absorb water.

Green roofs, for example, are designed to absorb large quantities of rainwater and so are rain gardens. Impermeable asphalt roads, parking lots, and sidewalks can be replaced with permeable, pervious, or porous pavement. Permeable surfaces allow rain to seep into the ground. This has two beneficial effects. First, it slows the passage of the rain into the receiving body of water. Second, the ground cleanses some of the pollution the rain has picked up while falling through polluted air. Third, it promotes the recharge of shallow groundwater that ensures that streams continue to flow even in the dry season. Fourth, it provides resiliency in the face of extreme weather events anticipated as a result of climate change.

Before going much further, we need to point out that unlike improvements to wastewater treatment plants or other such projects—all of which are on public property, many, if not most, stormwater projects will be on private property. Some of the properties will be on huge upscale shopping centers with large parking lots paved with asphalt. Some of them will also be on poor churches with large parking lots paved with asphalt. The point here is that neither the owner of the shopping center nor the pastor of the church is going to undertake stormwater reduction projects out of the goodness of their hearts. They will need to be compensated for their costs, if you want them into the game.

Now, as you can see, dealing with stormwater is not much like dealing with municipal or industrial sewage. There are no ratepayers with water meters. So, where does the money come from to deal with stormwater problems?

In December of 2014, EPA published a booklet entitled *Getting to Green: Paying for Green Infrastructure, Financing Options, and Resources for Local Decision-Makers.*

In terms of spending money on stormwater projects, the booklet suggests our friend, the 800-Pound Gorilla, the municipal bond market. It also suggests low-interest loans. Low-interest loans don't come from banks or anywhere else in the private sector, so that must refer to the Clean Water State Resolving Fund (CWSRF) or similar state programs. The CWSRF can provide excellent financing for green infrastructure projects. They can offer terms of up to 30 years. They can finance at market rates, which today are about 3.5% easily. Moreover, they have the authority to offer subsidized interest rates all the way down to 0%. So, the CWSRF is a truly excellent resource for financing green infrastructure projects.

So, spending money on stormwater projects is easy enough. However, both bonds and loans must be repaid. The money to repay has to come from somewhere, and it is this "somewhere" that has proven difficult so far.

The booklet suggests three possible sources of revenue for repayment: Revenue from (1) general taxes, (2) development fees and other charges for permits, inspections, and so on, and (3) stormwater-related revenues.

Using the State of Maryland as an example, the magnitude of the stormwater remediation problem is about $5.3 billion. This is about $1,000 for every man, woman, and child in the state. Suffice it to say that Maryland isn't going to raise $5.3 billion—or anything even close to that—from developer/development fees. Good idea. Yes, should they be imposed? Yes. But will they pay all the bills? No.

Alternative #2 is stormwater-related fees or taxes.

In the last 10 years, stormwater-related fees have become popular among local environmental officials. Depending on specific state laws, these fees can be imposed by a stormwater utility, if there is one, or by a unit of local government. In Maryland, it was imposed by nine specific counties plus the City of Baltimore.

The operative regulatory mechanism for stormwater is a Municipal Separate Storm Sewer System, or "MS4," permit. So, Maryland issued MS4 permits to its nine largest counties plus the City of Baltimore. It then passed a state law requiring each of these jurisdictions to come up with a plan to deal with their stormwater problems "and to fund these efforts with a new, specific stormwater fee."

It didn't take the news media long before they dubbed these new stormwater fees as "rain taxes." In some places, as you will see, the term "tax" might not be legally correct, but for our purposes, we will join with the media and call all of these stormwater revenue initiatives "rain taxes." Good name.

As Maryland was going through these exercises, so were many jurisdictions throughout the United States, either voluntarily, or to comply with MS4 permits. The way many of these jurisdictions approached the problem was to adopt a "polluter pays" policy.

As you know from above, there is a high correlation between stormwater problems and the degree of permeability of the land over which the storms occur. So, many jurisdictions decided to base their rain tax on permeability. To do so, they hired firms to do two things. First, review aerial photographs of the entire jurisdiction and identify the permeable and impermeable surfaces. Second, come up with an average, or model, parcel of land, whose degree of impermeability could be used as a standard. These were generally single-family home lots. These model parcels were called something like an "Equivalent Dwelling Unit" (EDU). They were then able to use these EDUs to measure impermeability all over the jurisdiction. For example, a church or shopping center with a large parking lot might be 10.8 EDUs or some such.

Now, having done all this, two issues arose. First was the "polluter pays" issue. Many jurisdictions decided to give property owners the option of paying the full rain tax for their parcel or to undertake projects to reduce the square footage of impermeable surface on their property. If they reduced the amount of impermeable surface on their property, they were rewarded with a reduction in the rain tax they had to pay. So, if you were a major "polluter," you paid the full tab, but if you reduced the pollution coming from your property, you would pay less rain tax.

Remember above, when I said that all of these stormwater revenues were—technically—like taxes? Well, the legal mavens point out that taxes are matters that you cannot avoid and have no choice except paying. On the other hand, you can opt out of fees. If you are willing to take a specific action, you don't have to pay the "fee," or at least, not all of it. So, in this case, if you opted to undertake a project to reduce the impermeable surface on your property, you could reduce your stormwater "fee."

This distinction between taxes and fees has some serious financial implications. When it comes to taxes that are associated with land, there are properties that are categorically exempt. Churches and government buildings are good examples. If you enact a true rain "tax," then government buildings and churches will not have to pay it, because they are exempt. On the other hand, if you enact a program where the churches and government buildings have the option to reduce their liability by taking specific actions, then they will be subject to the fee.

Churches aren't a big problem, but think, for example, of the counties that abut Washington, DC. There are many federal government buildings in these jurisdictions. Montgomery county, which is adjacent to Washington in Maryland, enacted a stormwater fee that property owners could reduce by taking action to reduce the amount of impermeable surface on their property. The federal government pays this stormwater fee to Montgomery County for all its buildings in the jurisdiction. (As of this writing, however, the State of Maryland has refused to pay Montgomery County for the state buildings in the jurisdiction.)

However, as you will see below, undertaking a project to reduce the impermeable surfaces on your property may be very unlikely; so, we will continue referring to the entire stormwater revenue issue as a rain tax.

The reason that undertaking an impermeable surface reduction project is unlikely has to do with the structure of the rain tax itself. Let us take the hypothetical case of a homeowner who has a 500 square foot driveway of impermeable asphalt on his 1 EDU parcel of land. Let us say that the rain tax on his EDU is $100 per year. Now, wanting to be a good citizen, the homeowner investigates replacing the impermeable asphalt with permeable pavement. If he does so, the county tells him that his rain tax will be reduced from $100 to $20 a year. So, the homeowner contacts a couple of qualified contractors and gets estimates. The estimates are about $10 per square foot, or $5,000. The homeowner then goes to his local bank, which offers him a $5,000 home improvement loan at an interest rate of 7% and a term of 7 years. The homeowner then realizes that his annual payment to the bank will be $928. So, he is confronted with the choice of paying a rain tax of $20 per year PLUS $928 to the bank for his impermeable surface reduction project—"or continuing to pay the full $100 annual rain tax." Which would you do?

So, it is not economical to opt out of the full rain tax. Therefore, it looks much more like a tax than a fee. However, as long as the option is there, it's legally a fee.

The second issue with imposing rain taxes is the amount of the tax itself. How much is the county going to charge per EDU?

Now, one would intuitively think that the amount of the rain tax would be the total annual cost of the program—including the debt service payments on any bonds or loans—divided by the total number of EDUs. Well, that's the theory. However, there's generally a gap between theory and practice.

I can illustrate this gap with some firsthand information that I gained while Maryland's 10 jurisdictions were going through the throes of creating their own rain tax regimes.

During that time, I had occasion to visit with the responsible government officials in one of Maryland's wealthiest counties. I asked what numbers they were looking at for the rain tax implementation ordinance that they were preparing.

The answer was that the engineers were talking $150 per EDU, but the members of the County Council—the politicians—were talking $50 per EDU. I was totally incredulous. I had done the numbers before the meeting. I had a very good idea of the annual cost of the program that the county needed and I knew how many EDUs they had come up with.

The real number—the annual cost of the program divided by the total number of EDUs—was a little more than $300 per EDU per year! $300 versus $150 versus $50! The staff had lowballed the number to avoid the wrath of the press and the politicians and then the politicians had lowballed the number again to escape the wrath of the news media and the voters. The end result was a rain tax a little more than what the politicians originally wanted but far, far less than the money needed to get the job done.

So, for several reasons, rain taxes can be fraught with difficulties both for implementing and for raising enough revenue to get the job done.

This takes us back to the first revenue alternative in the EPA booklet: a general tax or revenue stream.

For most units of general government, such as cities, towns, and counties, their major sources of revenue are real property taxes and sales taxes. The key issue here is the breadth and scope of these revenue sources. Very few properties—churches, government buildings, hospitals, and most other not-for-profits—are exempt. Most properties are fully taxable.

Likewise, it is with the sales tax. Depending on the state, very few articles of merchandise—may be food—are exempt. Again, churches, not-for-profits, and the like usually don't pay sales tax. However, again, exempt items and exempt organizations are a very small part of the population. In general, sales taxes and real property taxes are huge, extremely broad sources of revenue.

With water and wastewaters, it is the long history of full and prompt payment by the ratepayers that is the key to their strong credit ratings. With local governments, it is the size and the breadth of their general tax collections that are the source of their credit quality.

As you know, this credit quality is key to financing projects at the lowest possible interest rates and for the longest possible terms.

So, what about stormwater? Can stormwater projects be paid for out of general revenues, or can the debt service on bonds and loans issued for stormwater projects be paid for out of these general revenues? The answer is: if necessary, yes. However, there are better ways of doing things.

When it comes to environmental projects, what you call things take on an aura of great importance.

The best example of this that I know of comes, again, from Maryland. In 2004, the state got serious about requiring their sewage treatment plants to implement Enhanced Nutrient Removal. This reduced the nitrogen from sewage treatment plants from discharging effluent that contained 8 parts per million (ppm) to 3 ppm. This is very expensive. The rule of thumb in wastewater is that costs rise exponentially as treatment levels are increased. The state did not want to hear whining and complaints from their wastewater utilities about how high the costs were and how much they would have to raise rates. In addition, they didn't want the utilities to drag their feet, and so, in a stroke of brilliance, Governor Bob Ehrlich, a Republican, and the Maryland General Assembly, with a large Democratic majority, just took the bull by the horns. They imposed a charge on every sewer connection in the state as well as on every septic system. The charge was $2.50 a month or $30 per year for a residence. For office buildings and other commercial structures, they created almost an EDU-like system. If the average residence had 2.5 toilets, that was the base. For an office building or shopping mall with 50,000 square feet, they estimated *x* number of toilets and charged accordingly. So, if you owned an office building, you might get a bill for $25 a month, which represented 10*x* the number of toilets in an average household.

Needless to say, the news media had a field day with concept. The politicians called it the "Bay Restoration Fee." (It wasn't a fee; it was a tax. There was no way to avoid it. You couldn't very well disconnect your home from the sewer system. However, Maryland has gotten away with calling it a "fee"; so, churches and government buildings must pay—just like they pay for sewer services.) The news media, of course, called it the "flush tax." What a wonderful name!

More importantly, however, what an ingenious concept! Note how broad it is. Every residence and every other building—including churches and not-for-profits, pay the flush tax. Even homes with septic systems pay. These systems are not connected to any sewage treatment plant, obviously, but they still pay.

Even more incredibly, the flush tax has been so successful that, in 2012, they doubled it to $5 per month! Wait! How could this happen? Where was the massive public outcry?

Well, there was some opposition, as there is to everything in politics, but it wasn't much.

The reason is what's in the name. People may not know the exact official name of the tax—the Bay Restoration Fee, but they know that the money is going to improve the Chesapeake Bay, which they approve of, wholeheartedly. Even the name "flush tax"—funny as it is—clearly reminds them of the purpose of their payments.

This concept is critical. Maryland raises about $120 million a year now from its flush tax. It could also have simply added $120 million to the state sales tax or real property tax, but no, the political leadership had the good sense to realize that public support would be generated—and opposition be blunted—by associating the income directly with the peoples' beloved Chesapeake Bay. And they were right!

So, what is the best way—from a financial perspective—to raise money for stormwater reduction programs and projects?

Well, when thinking about a stormwater fee, there are two countervailing considerations. First, since it is a fee, there will be more EDUs to spread the cost over, because no properties are exempt. So, the stormwater fee will be lower. Second, and on the other hand, if the stormwater fee is very low, then—per the example above of the homeowner with the impervious driveway—no one will opt out. This will mean that it will be the stormwater program, itself, that will have to pay to have the homeowner's driveway repaved with permeable material. This is not necessarily a bad thing. Remember that the homeowner's bank was going to charge him $928 a year. Well, if the stormwater program undertook the same project, the homeowner could finance it for 30 years at a rate of, say, 3.5%, for an annual payment of $272. So, since the homeowner is paying only $100, it means that the rest of us will have to pay the difference. However, this is an issue that the politicians—just like the ones in the wealthy Maryland County—will ultimately have to deal with.

So, that is the story with stormwater fees with opt-out provisions.

The other alternative is a straight rain tax with no opt-out provisions. Again, you need to do the math. Since there are exemptions, how much higher will the tax have to be because of the exemptions?

The final consideration with a straight, general tax is: what will be its basis? Will it be charged by parcel, regardless of size? Will it be charged based on the square footage in each parcel? Will it be *ad valorem*, that is, based on the assessed value of each parcel? Or—and it certainly can be—will it still be based on the amount of impermeable surface on each parcel?

So, as you can see, there are many options as to how funds can be raised to pay for stormwater programs.

So, the point here is that when it comes to finding revenue to pay for stormwater projects, the source or the sources of the money must be large and broad. This is true whether they are "opt out-able" fee programs or straight taxes. Moreover, in order generate public support and mute opposition, they should be closely identified with the health and cleanliness of the lakes and rivers and bays that they protect.

17 Groundwater Source Protection and Watershed Preservation

Among the nonpoint source projects eligible for funding under the Clean Water State Revolving Fund (CWSRF) are watershed protection projects. For example, this is how the State of Illinois describes its program: "The program includes providing funding to these groups to implement projects that utilize cost-effective best management practices (BMPs) on a watershed scale. Projects may include structural BMPs such as detention basins and filter strips, nonstructural BMPs such as construction erosion control ordinances and setback zones to protect community water supply wells. Technical assistance and information/education programs are also eligible."

As you already know, some 96% of the $111 billion of financial assistance provided by the CWSRF has gone to sewage treatment plants. However, as you also know, that game is changing. There are two other categories of eligibility under the Clean Water Act. The first is for nonpoint source pollution control projects. These are governed by Section 319 of the Act. The other eligible category includes projects for estuary protection. These are governed by Section 320 of the Act and are dealt with in the chapter of this book on the CWSRF.

SECTION 319 AND INTENDED USE PLANS

Section 319 of the Clean Water Act addresses nonpoint source pollution. Financing of nonpoint source pollution projects was dealt with in Chapter 15 of this book. Here, we will focus on the nonpoint source projects that deal with watershed preservation.

Section 319 requires the states to inventory their receiving waters and identify those water bodies that are impaired by nonpoint source pollution. The states are then required to prepare what is known as a "319 Program," which describes how the state will deal with these nonpoint sources of pollution. These 319 Programs are sent to EPA for approval.

Protecting watershed is an eligible activity under section 319. In order to finance such projects, this activity should be spelled out in each state's 319 Program. That's step one in getting CWSRF help.

Title VI of the Clean Water Act sets forth the provisions of the CWSRF. As might be expected, there are eligibility categories as well as procedures for this program too.

Title VI requires states to create Intended Use Plans (IUPs) that set forth the types of projects on which the CWSRF intends to spend its funds. Each year, each state must prepare a draft IUP, put it out for public comment, and submit it to EPA. Once the IUP is approved, the states then solicit applications for funding. Getting watershed protection into the CWSRF's IUP is step two in getting its financial assistance. Almost all of the individual state CWSRFs list groundwater source protection or watershed preservation on both their 319 Programs and their IUPs. Notably, Minnesota does not, but that's because it has a state fund that can be used directly for those purposes.

CLIMBING THE PROJECT PRIORITY LIST OR NOT?

Title VI also requires states to prepare Project Priority Lists (PPLs), which actually list the projects to be funded in the order in which they will be funded and in the amount in which they will be funded.

For point source projects, each project application is scored under the rubrics created by each state for ranking its own project. States have a finite amount of money to lend; therefore, in times of strong demand, there is an imaginary line on each PPL. If your project is above the line, it gets funded; if it is below the line, it doesn't get funded unless a project with a higher priority is cancelled or postponed and your project moves up the list to where it is above the "line."

So, how, in general, do watershed protection projects stack up in this prioritization game? Answer is: not well. A publicly owned treatment work (POTW) that cleans X gallons of water a day and is going from secondary to tertiary treatment will remove Y pounds of nitrogen, for sure. How much nitrogen will be prevented from entering receiving waters if an acre of forest is preserved rather than turned into farmland or a housing development with large over-fertilized lawns? Who knows? At least, that's the way most engineers look at these matters. Although our understanding of how natural infrastructure filters nutrients and other substances in water is constantly improving, we have ways to go with regard to measuring impacts. It is, therefore, no wonder that the CWSRFs have spent 96% of their funds on POTWs.

There is some good news, however: individual nonpoint source projects do not have to be named and they compete individually against point source projects for CWSRF dollars. Nonpoint source pollution control projects are eligible by category, as long as the category is listed in the 319 Program and the IUP. However, the bad news is that they still have to compete for often-scarce CWSRF dollars. How to address this challenge?

GUARANTYING RESULTS

Despite the challenge of quantifying their filtering benefits, source water protection programs offer many benefits to the water community. How then can watershed protection projects navigate the political/prioritization thicket of the CWSRFs and get funded? The answer is to seek a loan guaranty from the CWSRF and not a direct loan. This is step three in the CWSRF financing procedure.

As you know from Chapter 10, the loan guaranty capacity of the 51 CWSRFs collectively is estimated between $1 *trillion* and $3 *trillion*. That's trillion, not billion. In the summer of 2015, the New York State Environmental Facilities Corporation estimated that it had over $100 *billion* of financial guaranty capacity. That's a conservative estimate.

So, the point is that although a CWSRF in any particular state might not have available fund to finance all its point source *and* nonpoint source customers like watershed preservation, they most certainly will have the capacity to guaranty it.

What does this mean? It means that with a guaranty from your state CWSRF, you will be able to issue a bond at AAA rates, which are the lowest rates available. (All CWSRFs are rated AAA.) You will also be able to issue for long terms. Land lasts forever, so you will be able to issue a 30-year bond with the CWSRF guaranty. Having the longest terms and the lowest rates will reduce costs and enable more projects to get done.

Title VI of the Clean Water Act authorizes both direct loans and loan guaranties. Of over 30,000 projects that the CWSRFs have funded since 1987, fewer than a dozen have involved the loan guaranty authority. The reason for this is that the direct loans are made at subsidized interest rates, which all of the local POTWs want.

In general, subsidized CWSRF loans are made at rates that are about 50% of market rates. In today's world, that's about 2%. Vermont makes 0% loans. This is the money that is subject to the often-fierce competition on the PPLs. Meanwhile, projects seldom seek guaranties.

Does using the guaranty authority in any way limit the amount of subsidized loans that a state can make? Functionally, no.

Although the guaranty authority under the Clean Water Act has not yet been used for watershed preservation projects, there have been some successful projects that have been funded through one of the major financial mechanisms used by the individual state's SRFs for nonpoint source projects: linked deposits or "loans-to-lenders" programs. This funding method is described more detail in Chapter 15.

There was an excellent example of a watershed preservation project funded by a linked deposit in Annapolis, Maryland.

The project involved the acquisition in 2002 of about 110 acres of undeveloped woodlands and beaches located in the Bay Ridge Subdivision on the southern edge of the City of Annapolis. This constitutes approximately one-third of the land area of the entire community.

The project sponsor was the Bay Ridge Civic Association ("the Association"). The Association consists of the homeowners in the subdivision. There are about 425 eligible properties in the community and about 360 active members of the Association. The owner/developer agreed to the sale of the acreage to the Association.

The all-in project cost, including the purchase price and some tax payments, was $5.5 million. Of this amount, the Association raised about $1.2 million from donations from its own members. It also sold 6 lots to adjacent owners (who would not develop the land but use it for a buffer) and 4 lots to new owners, who built homes on the lots. The total realized from these sales was about $1.5 million. The Association also obtained a "Program Open Space" grant for $450,000 from the State of Maryland and $100,000 from additional, other sources. The Association borrowed the remaining $2.25 million from the CWSRF, which is managed by the Maryland Water Quality Financing Administration, a division of the Maryland Department of the Environment.

The primary reason the Association purchased the woodlands was to prevent them from being developed, to conserve an important natural resource, and to preserve a closely knit community. Because the woodlands were on the shore of the

Chesapeake Bay and its tributaries, the project had a sufficient positive impact on Bay water quality to qualify for a CWSRF loan.

Now, Maryland has a Special Taxing District law, which enables counties to create Special Taxing Districts. The Association petitioned Anne Arundel County to amend the purposes of an existing Special Community Benefits District to include the acquisition and maintenance of real property and the repayment of any loan. The Association then requested a $2.25 million loan from the CWSRF. It was the CWSRF's policy to effect nonpoint source pollution loans through a "loans-to-lenders" program with local banks. Loans-to-lenders programs typically involve the CWSRF placing its funds—in this case, $2.25 million—on deposit with the local bank at a reduced rate of interest. So, although, if at any given time, the bank was offering certificates of deposits (CDs) for, say, 4%, the CWSRF would agree to accept, say, 2% if the bank were willing to reduce its interest rate to the project sponsor/borrower by the same amount. (It should be noted that with loans-to-lenders programs, it is the bank—not the CWSRF—that bears the risk of loss in the event of nonpayment.)

The Association approached the Columbia Bank, which was willing to make a 20-year loan, with a 30-year amortization schedule to the Association at a rate of 6.975%. The Bank was offering 20-year CDs at 4.975%. The CWSRF agreed to accept 1% interest rate on its loan in return for the Bank's reducing the interest rate on the loan to the Association from 6.975% to 3%.

The principal reason the bank was willing to make this loan was the secure nature of the cash flow available to service the loan payments. The Bay Ridge Special Community Benefits District is administered by the Board of Directors of the Association. The Board votes to impose the annual loan payments as a tax on each property in the District. The Maryland Special Tax District Act provides that any charges imposed on properties in such districts can be collected by the respective counties as part of their real property tax collection process. Furthermore, such payments constitute liens on the respective properties. So, the repayment stream to the Bank was highly secure.

To summarize, the Association raised the $5.5 million needed for the project as follows: $2.25 million from the CWSRF, $1.5 million from the sale of lots, $1.2 million from donations from Association members, $450,000 from a "Project Open Space" grant from the State of Maryland, and $100,000 from other miscellaneous sources.

There are several brilliant elements to this transaction. The first was the use of a 30-year amortization to reduce the annual cost of the loan. The second was the securing of the loan with liens on the real property owned by the members of the Association. The third was the CWSRF's flexibility in being willing to accept these highly unique aspects of the transaction.

HOW MUCH MORE IS AVAILABLE?

The net assets of the CWSRFs can be used as a short-hand measure of their guaranty capacity, much in the same way that the common stock of a bank can be used to measure its capacity for making loans. If a bank has $10 billion of stock, you should expect to see its loan portfolio in the $80–$120 billion range. In short, the ratio of loans to stock is 8–12 : 1. In this regard, the combined net assets of the CWSRFs are over $40 billion. However, according to Standard & Poor's (S&P) and the other international credit rating agencies, wastewater loans are much safer than the commercial and consumer loans that banks make. So, in this case, you would expect to see 25+ : 1 ratios of guaranties to net assets (S&P says 75 : 1). However, at a ratio of only 25 : 1, the combined CWSRFs have $1+ *trillion* of guaranty capacity. At present, the total leverage ratio of all the CWSRFs is a little over 2 : 1, which means outstanding direct loans in the $80–$100 billion range. Thus, the combined CWSRFs have well over $900 billion of unused financial capacity. In short, the CWSRFs cannot make an infinite amount of guaranties, but their unused financial capacity is so great that there is no real functional limit.

So, the CWSRF can be used to guaranty bonds issued by local governments or even not-for-profit corporations for watershed protection projects.

SATISFYING THE CWSRF'S RULES

Regardless, however, of whether a project takes advantage of a direct loan, a linked deposit, or a guaranty from a CWSRF, it must obey the other CWSRF rules. Most of these rules pose no difficulty for typical watershed protection projects. That said, most watershed protection projects are paid for with grants or donations—money that does not need to be repaid. The CWSRF money, whether as loans or guaranties, needs to be repaid. Furthermore, one of the CWSRF rules requires a "dedicated revenue stream" for repayment.

A "dedicated revenue stream" has two characteristics. It must be reliable, and it must, of course, be dedicated. As you could see from the Annapolis example, the homeowners' fees—collected by the county—were a very, very reliable income stream.

By "reliable," bond lawyers mean sewer fees and charges that system users pay every month, quarter, or year. Taxes and other fees can be used as well. However, these fees and charges cannot be voluntary and they cannot fluctuate to any significant degree. This means that conservation nongovernmental organizations (NGOs) and other such organizations cannot use their dues revenue to meet the "dedicated revenue stream" test. Such organizations can, however, use endowment income to meet the test. This can be tricky. If the endowment has earned at least $10 million a year for the last 10 years, and it only needs $1 million to pledge against a CWSRF-guaranteed bond, the bond lawyers and rating agencies would probably let an organization get away with it as long as it pledges the "first $1 million of earnings" from the endowment. Likewise, segregating specific, highly-rated, fixed-income securities and pledging their interest earnings would also work. So, it can be tricky, indeed; however, it can be done.

The final caveat, of course, is that these revenues must be "dedicated." This dedication must be formal. It can be done by a local law or ordinance, in the case of a local government. Where businesses, NGOs, or individuals are concerned, such pledges can be secured by contract. However, a bond counsel's opinion will be needed on the contract.

Another promising source of dedicated revenue that could play a big role in loan guarantees for watershed protection are the source water or watershed protection fees and other financial structures being used in dozens of communities across the country. Raleigh, North Carolina, for example, charges its 600,000 system users $0.0748 per hundred cubic feet. This volumetric rate generates about $1.3 million annually, which is dedicated for watershed protection activities. This is a good boost to source water protection efforts, but imagine what could be accomplished if this dedicated revenue was used as a loan guaranty—it could generate almost $20 million, which could be used to implement projects at today's prices.

Organizing a "dedicated revenue stream" is the fourth and final step in the CWSRF financing process.

So, as you can see from above, there are direct loans, linked deposits, and financial guaranties that can be provided by the CWSRFs to finance watershed protection projects. If the competition for direct loans is too fierce and too political, then the financial guaranty route is the way to go. As of this writing, however, several states have reported insufficient demand for their direct loans to POTWs. So, it is important to find out just what the financial situation is with the CWSRF in your state in order to determine whether to apply for a direct loan, a linked deposit, or a financial guaranty.

So, in summary, the CWSRF can definitely be used to finance watershed protection projects, and with $1 trillion of financial capacity, it should definitely be used to do so.

18 Energy-Efficiency Projects

On August 13, 2013, the State of New York set an astonishing legal precedent for promoting energy-efficiency.

On that day, the New York State Energy Research and Development Authority (the "Authority") issued $23.4 million in bonds to fund residential energy-efficiency projects. The bonds were rated AAA, thanks to a financial guaranty from the New York State Environmental Facilities Corporation (EFC). What was astonishing was that the EFC, which manages the Clean Water State Revolving Fund (CWSRF) in New York, used legal authority under the Clean Water Act—not the Clean Air Act—to effectuate the guaranty! In effect, the EFC used the net assets of its CWSRF program to credit enhance the bonds of its sister agency that does energy projects. This is the first-ever transaction to provide linkage between clean energy and clean water programs.

The transaction was so significant that it won a National Deal of the Year Award from *The Bond Buyer*, the authoritative publication in the public finance industry.

In July 2013, both Moody's and Standard & Poor's rating services assigned their highest ratings to the Authority's residential energy-efficiency financing revenue bonds series 2013A, which were issued 3 weeks later. The ratings are based on the EFC's commitment and ability to make full and timely payments of principal and interest, should the Authority become unable to do so. This use of a guaranty by the EFC will significantly expand New York State's toolbox to fund residential and commercial energy-efficiency projects. This is an entirely new territory for the Environmental Protection Agency's (EPA) CWSRF.

As discussed in Chapter 11, Section 319 of the Clean Water Act is entitled "Non-point Source Management Programs." Although pollutants such as nitrous oxides may emanate from a smoke stack, exhaust pipe, or some other "point source" of *air pollution*, once they are airborne and destined for some receiving water body, they become a "nonpoint source" of *water pollution* as far as that receiving water body is concerned. As such, the atmospheric deposition of air pollutants into a state's water bodies may, or may not, be included in any given state's Non-point Source Management Program (319 Program). In the case of New York, "atmospheric deposition" is listed prominently on page one of its 319 Program.

How exactly did New York set this groundbreaking and important precedent? New York did it through an excellent marshaling of the facts and by a close dialog with the SRF program management at the EPA in DC.

New York had established that "(b)urning fossil fuel to generate heat and electricity in NYS contributes to atmospheric deposition of air pollutants into NYS water bodies. NY's Non-point Source Management Program identifies atmospheric deposition from fossil fuel combustion as a significant source of water quality impairment and calls for additional controls over, and reductions in atmospheric deposition of such air pollutants into NY's waters." EPA agreed.

So, what does this all mean?

It means that New York has used a clean water finance program to facilitate the funding of clean energy projects. More importantly, it means that other states can do the same.

Before we discuss how other states might take advantage of this opportunity, we need to analyze the New York transaction.

Under Title VI of the Clean Water Act, states are given several options to finance clean water projects. They can make direct loans. They can "buy or refinance the debt obligation of municipalities and intermunicipal and interstate agencies within the State." In addition, most importantly, the Act authorizes SRFs "to guarantee, or purchase insurance for, local obligations where such action would improve credit market access or reduce interest rates." This is the provision under which the EFC guarantied the Authority's bond issue.

Since 1987, the SRFs have provided over $110 billion of financial assistance for clean water projects on a national basis. This has involved over 36,000 transactions. That said, only one—one out of 33,000—has involved the use of the guaranty authority in the capital markets. So, New York's amazing precedent was actually two equally amazing precedents: first, the use of a clean water program to finance an energy-efficiency program, and second, the first-time use of the guaranty authority in the capital markets!

The New York's precedent means that in states where air deposition is a significant contributor to water pollution, their SRFs can provide financial assistance to residential, non-governmental organizations, and small-business energy-efficiency projects. Theoretically, such authority could also be used to guaranty bonds issued to finance pollution reduction systems at major power plants; however, individual state policies may prohibit such action. In any case, the Clean Water Act can be used to reduce air pollution and energy consumption.

Will these new precedent-setting uses of SRF funding capacity result in fewer dollars for traditional sewage and other clean water projects? The answer is a resounding "no." As discussed in several places in this book, the combined 51 SRFs have over $1 *trillion* of financial guaranty capacity. So, venturing into new ground such as this won't even dent their fender.

So, how can other states avail themselves of this same opportunity? The four steps for accomplishing this are

1. Identify whether atmospheric deposition is a source of water pollution in the state. The place to find out about this is the National Atmospheric Deposition Program (NADP), which is an EPA-sponsored program housed at the University of Illinois. The NADP website is loaded with information about deposition conditions in every state.
2. If so, include it in the state's 319 Program. A sample glance at the 319 Programs of the 10 states with the largest populations yields some interesting results. New York, Pennsylvania, North Carolina, and Illinois mention atmospheric deposition either on page one or otherwise prominently in their

319 Program descriptions. California, Texas, Florida, Georgia, and Ohio don't mention it at all. Michigan has a very surprising provision that identifies air deposition as a significant source of water pollution but says that they will use the state's regulatory powers to reduce it.

3. Once it is in the 319 Program, then include "activities to reduce atmospheric deposition" in the SRF's Intended Use Plan (IUP).
4. Once this language is in the IUP, you are good to go. Applicants may apply to the SRF for a guaranty for energy-efficiency projects.

Why is all this necessary? It looks like a lot of work.

It is a lot of work, but it's worth it.

Promoting energy efficiency is certainly a worthy goal. That said, left to the private sector, here's what happens:

A homeowner wanting to, say, insulate his home or replace leaky doors and windows at a cost of, say, $20,000 would most likely go to his local bank for a second mortgage or simple home improvement loan. The bank would likely offer something like a 7% rate and a 7-year term. That would result in a monthly payment of $302. If the homeowner moved away the next year, he would have to pay off the balance.

Since home insulation and new doors and windows have a service life of 30 years, the SRF can offer 30-year term financing. Furthermore, at this writing, SRF market (not subsidized) rates are about 3.5%. So, financing with an SRF guaranty would result in a monthly payment of $91.

Now, put yourself in the position of a state energy policy maker that wants to promote energy-efficiency. Don't you think more people would be willing to do projects that cost $91 than projects that cost $302? Hello? We don't need to be rocket scientists to figure this one out!

Two final comments:

First, the CWSRF has done over 36,000 transactions, totaling about $110 billion. This means that their average-sized transaction is over $3 million. That is what the CWSRFs have historically done—make $3 million loans to sewage treatment plants. So, what do you think will happen when a homeowner walks in the door, asking for a $20,000 loan for home insulation?

The problem here is "aggregation." The SRF staffs can't deal with hundreds, or thousands, of individual requests for relatively small loans, or, in this case, loan guaranties. The SRFs need another entity to aggregate these loans—just like the Authority did in New York. The SRF can deal with a guaranty of a portfolio of loans—as in New York—but not individual ones. So, who will be the aggregator? It will probably be different in each state. However, in your state, it could be you! As you will see below, the energy-efficiency loan should look—legally—as much as possible like a water or sewer charge.

The second comment relates to the statement above that if a homeowner went to his local bank and then moved away the following year, he would have to pay off the loan. With an SRF program, this is not necessarily so.

Energy-efficiency loans guaranteed by an SRF should be organized through the local government's real property tax collection system, as described in Chapter 9 (Credit Enhancement Techniques). There are two benefits to this strategy. First, it will make the loan repayment stream highly secure. The SRF people will not have to sit up all night biting their nails and worrying about whether their energy-efficiency borrowers will repay their loans or not. These energy-efficiency loans should look as much as is humanly possible like water or sewer charges.

Second—and most importantly for the viability and salability of the whole energy-efficiency program—if the homeowner moves away next year, he "does not" pay off the balance of the loan. Instead, it remains as a lien against the property on the books of the local government and the new homeowner—who gets the full benefit of the insulation and of the new windows and doors—very fairly gets the pleasure of paying for them. (It should be noted that there is still some controversy about the legal priority of liens like these. This was all caused by the Federal Housing Finance Agency that regulates Fannie Mae, the largest purchaser of home mortgages in the country. However, these issues are slowly resolving themselves in each state.)

It's been almost three years since the New York transaction, and no other state, has yet replicated it. Why not? Well, there are several reasons. As mentioned above, aggregation is a huge problem. Second, there is a major disconnect between CWSRF staffs and the staffs of state energy offices. They literally know little or nothing about each other's programs. Finally, there is the question of the guaranty itself. As noted above, it has been used only once in the capital markets—by New York. The CWSRF staffs just aren't familiar with guaranties.

That's the bad news. The good news is that Pennsylvania is getting into the scrum, and it is PennVest, which manages the CWSRF that is providing the financial support. However, PennVest is not using its guaranty authority; rather, it is using its "investment" authority under a different section of the Clean Water Act. All CWSRFs have the authority to invest funds. However, most investments have been temporary—between the time an SRF receives a loan payment and the time when it uses these funds to make another loan to a different borrower. However, the Clean Water Act does not limit "investments" to these temporary facilities.

So, here's how the Pennsylvania program works. The program is called "Keystone HELP" (the "Program"). (Remember that Pennsylvania is the "Keystone State" because of its shape.) The "HELP" stands for "Home Energy Loan Program." The Program is a Public–Private Partnership between the Pennsylvania Treasury, PennVest, Renew Financial, and local governments.

Interested homeowners can either call the KeystoneHELP hotline or contact an enrolled contractor directly. If necessary, the program can help homeowners find a qualified contractor. Once that connection is made, the homeowner works with the contractor to select from a set of pre-approved energy and water conservation measures. From single-measure HVAC replacement to electric vehicle charging and whole home retrofits, hundreds of technologies qualify.

The homeowner then applies for KeystoneHELP financing by phone or online. This process is designed to be fast and easy, since most contractors won't use financing that requires extra paperwork or places an extra burden on the homeowner.

Every homeowner who qualifies for KeystoneHELP financing benefits from the same low interest rate, currently set at 7.99%. This is achieved on a loan-by-loan basis, depending on the credit score of each individual borrower. For all borrowers, PennVest funds occupy a first loss position in the capital stack, with private capital in a senior position. For borrowers with high credit scores, the PennVest capital commitment is relatively small. For borrowers with lower credit scores, the PennVest capital commitment is much greater. At a portfolio level, for every one dollar that PennVest commits to energy-efficiency loans in Pennsylvania, four dollars in private capital are committed by institutional investors. For its role in the program, PennVest receives a low single-digit return, consistent with its programmatic objectives.

Refer to Chapter 9 discussion of "Tranching" as credit enhancement mechanisms. Note that PennVest's "first loss" position in the financing is in fact a "Z Tranche." Since PennVest is a government agency and not a private equity fund, it receives a very modest rate of interest for its critical support of the Program.

As we go to press, Pennsylvania has not done a KeystoneHELP financing with PennVest's support, but one is planned for the near future. Moreover, there is no reason why such a structure wouldn't work.

19 Resiliency Projects

This chapter discusses how coastal resiliency projects can be financed. It breaks down coastal resiliency projects into their two logical types and then discusses various financing options for each type of project. Among the financial issues discussed will be Public–Private Partnerships, a System Benefit Charge (SBC) Model, Special Districts, Seasonal Charges, and Geographically Targeted Charges. The concept involved in both the Seasonal Charges and the Geographically Targeted Charges is to draw revenues for coastal resiliency from those who use and enjoy the coasts and who, otherwise, don't contribute to the well-being of the coasts.

As you will see, this chapter deals primarily with public, or community, facilities. It does deal with private businesses and residences, but only in a collective sense. As far as individual homes or businesses are concerned, there is readily available financing for hardening projects through the Clean Water State Revolving Fund (CWSRF). In 2014, Congress enacted several amendments to the Clean Water Act. One of these changes specifically authorized the CWSRFs to finance resiliency projects on private property. So, if you live on the beach and pay exorbitant flood insurance premiums, and need to raise your home on pillars, or stilts, you can get financing through the CWSRF if you wish. You may not get subsidized interest rates, but you will certainly get market rates, which today are a little over 3%. Moreover, you will certainly get a 30-year term for such projects, which, as you know from reading Chapter 8, is the most important aspect of financing. Finally, by undertaking such a project, you will most certainly make a substantial reduction in your flood insurance premiums.

Coastal resiliency projects can be divided into two groups. The first group involves beach replenishment, dune building, and back-bay and small-inlet dredging. These projects protect coastal communities from serious storm damage. We can call this group "Disaster Mitigation Projects."

"Post-Disaster Resiliency" projects are a second and very specific category. This category involves the reconstruction of *only* the "core economic drivers" (CEDs) of a community. For example, many coastal communities rely on tourism for their economic existence. Therefore, CEDs in that community are those specific amenities that attract tourists. Beaches are CEDs and so are boardwalks, the "beachy" shops on the boardwalks, amusement parks, and other entertainment facilities.

Providing for the rebuilding of CEDs is not a general replacement for flood insurance. Rather, it is a replacement program for the CEDs in only the coastal communities that are destroyed by hurricanes or other extreme weather events. Using Ocean City, Maryland, as an example, Post-Disaster Resiliency Projects would include reconstruction of the beach, reconstruction of the boardwalk, reconstruction of the stores on the boardwalk, and reconstruction of recreational facilities such as the amusement parks. In short, the CEDs are the facilities that lure visitors to the coastal community. They are the facilities that need to be rebuilt immediately after a storm—with no delays and no hassles—in order to get the community's economy going again.

Below are some alternative means of financing these projects.

SYSTEM BENEFIT CHARGE MODEL

This model addresses three essential strategies for protecting oceanside communities: beach replenishment, dune rebuilding, and channel dredging. Strong beach structures protect communities from storm surges and high waves. The same is true of dunes, which protect inland property. Dredging has a critical role in flood mitigation.

This model presents one possible mechanism for financing pre-hurricane fortification programs to replenish beaches, rebuild dunes, and dredge critical channels. One assumption underlying these concepts is that these programs must be carried out every 4 years. This means that 25% of the money necessary should be raised each year. It also means that the use of long-term municipal bonds is not feasible.

The central concept behind this finance mechanism is that the fortification programs protect the economies of beachfront communities. What good is it to harden water, sewer, cables, gas, and electric utilities if the businesses that use these services are destroyed? No businesses, no jobs; no jobs, no communities.

Gas and electric utilities, together with state public utility commissions, have taken the lead in making their facilities resilient. Many states have added SBCs to the electricity bills of consumers. These charges were originally designed to provide funds for utilities to assist their residential customers in installing energy-efficiency measures such as home insulation, modern HVAC, and modern electric appliances. However, lately, many electric utilities have applied to their respective state public utility boards for permission to use these funds for hardening, or resiliency, programs. In New York, ConEdison has mounted an over $1 billion hardening program in the five boroughs of New York City alone. This program is financed by SBCs.

This existing SBC/hardening concept can be expanded to pay for fortification strategies, especially beach replenishment and dune replacement, which will protect entire communities from hurricanes. In a sense, it will be an extension of each utility's own hardening program.

In short, each utility will be asked to pay a share of the quadrennial cost of implementing these fortification programs, a charge that will be passed along to their individual users according to their individual use.

Here is how such a system might work:

(Wherever possible, data mentioned below have been taken from the State of Maryland and/or Ocean City, Maryland, or Worcester County, where Ocean City resides.)

Water and sewer: A 2010 survey by the American Water Works Association done by Raftelis Financial Consulting, Charlotte, North Carolina, reported that the average combined water and sewer bill for households was 1.5% of median household income (MHI). In Maryland, the MHI is $69,272. In Worcester County, it is $55,487. So, the combined water/sewer charges for homes in Worcester County are, putatively, around $830 per year.

Electricity: According to the Energy Information Administration (EIA), the average residential electricity bill in Maryland in 2012 was $129 per month, or $1548 per year. The EIA has no breakdown by county.

Natural gas: According to the American Gas Association, the average annual residential gas bill in Maryland in 2012 was $760.

Telephone: According to the Federal Communications Comission, the average charge for a basic landline is about $30 per month. With state and local taxes and charges, this comes to about $50 per month, or $600 per year.

Cable TV: According to "Forbes", the national average cable bill in 2011 was $86 per month and is expected to rise to $123 by next year. So, let us use $115 per month, or $1380 per year.

In summary, here's what these charges look like on an annual basis:

Water and sewer	$830	16%
Electricity	$1548	30%
Natural gas	$760	15%
Telephone	$600	12%
Cable TV	$1380	27%
Total	$5118	100%

Now, let us say (again putting volumetric considerations aside) that the average household, as described above vis-à-vis its utility usage, needs to pay its share of $400 for the fortification programs. On an annual basis, this would be $100.

So, in this case, the $100 annual charge would be distributed across the five utility bills as follows:

Charges	Percentage (%)	Year ($)	Month ($)
Water and sewer	16	16	1.34
Electricity	30	30	2.50
Natural gas	15	15	1.25
Telephone	12	12	1
Cable TV	27	27	2.25
Total	100	100	8.33

As you can see, what might be called the "SBC Approach" does have the benefit of spreading out these costs over several revenue sources, making paying for them less painful.

PUBLIC–PRIVATE PARTNERSHIPS

As you know well from Chapter 14, a Public–Private Partnership (P3) isn't a finance mechanism; it is simply a structure for implementing projects. For the purposes of this chapter, one might think of a P3 responsible for rebuilding the CED of a community. In this case, the cost of the financing involved in such a P3 becomes critical.

"The financings for P3s depend on which of the Ps is bringing the money to the party."

If the public P brings the money, then the financing will generally be done through the municipal bond market. This means that the lowest interest rates and the longest possible terms will be available. This means the lowest possible cost for the project.

If the private P brings the money, then it will depend on the private P's appetite for return on investment (ROI) and the term or exit strategy, that is, how fast it wants its money back. If the private P wants an ROI north of 15% after 5 years, the project will be very, very expensive. If the private P wants 10% in 10 years, the project will still be very expensive.

To determine whether a P3 might be useful in financing a project, we need to categorize projects into two basic classes.

Class A: This includes projects with no associated income and no realistic possibility of creating a dedicated income stream. Class A projects can be funded in only four ways: gifts from private donors; government grants; legal settlements, including compensatory payments; and general obligation bonds, in which case a unit of local government issuing the bond is the donor.

Class B: In terms of identifying funding sources, projects can be further divided by the amount of income they have. As you will see below, projects that are suitable for the (taxable or tax-exempt) municipal bond market provide the lowest interest rates and the longest terms. So, if all that much income, taxes, user charges, or fees are not involved, then bonds are a good financing source. If there is substantial project income available, then other financing sources can be explored. They are described below.

It is of note that P3s can fall into any category. If the private partner is a donor, then it's a Class A project. If the private partner is a private equity fund, then it's a Class B3 project.

Class B2 includes socially responsible investors (SRIs), many of which will accept a lower return on a project, with exemplary social value.

Class B3 includes average investors seeking higher returns and willing to accept higher risks. Private equity funds and hedge funds are examples of these types of investors. It should be noted that public pension funds, sovereign wealth funds, and the like—which one might think to be SRIs—aren't necessarily so. The California Public Employees Retirement System is the largest pension fund in the United States. It has $294 billion of assets, $31.3 billion, or a little over 10% of which, is invested in "private equity." Its benchmark return rate for its private equity investments for the fiscal year ending June 30 was 15.4%.

Here are the classes of projects with their associated financing sources along with their terms, rates, and ROIs.

Class	Funding Source	Term (Years)	Annual Rate of Return (%)	Annual Return (Cost in $) per $1,000
A	Donors/G.O. Bonds	n/a	0	n/a
B1	Bond Market	30+	<5	65.05
B2	SRI Funds/Investors	~20	5–10	117.46
B3	Private Equity/Hedge Funds	5–7	15–40	240.36 (low) 491.36 (high)

In summary, once we know the type of private partner who is financing the project, we can determine the annual cost of that project to those who must pay for it.

SPECIAL DISTRICTS

Special Districts, or, more properly, Special Taxing Districts (STDs), can be used where there is a discreet geographic hardening issue to be addressed.

Think of beachfront property. Think of an inlet bay. The beachfront property needs beaches to protect it from storms. These beaches must be replenished about every 4 years. The inlet bay needs to be dredged. Again, this must be done periodically.

The state (or county, if appropriate) could create an STD whose boundaries are the inlet or the oceanfront properties at risk. This district could then impose charges on the properties in the district to support beach replenishment or dredging, as the case may be. These charges would be collected by the county as part of the real property tax collection system. Although both dredging and beach replenishment need to be done too frequently to use long-term tax-exempt bonds, intermediate term (3–5 years) Revenue Anticipation Notes (RANs) can be issued by the STD to pay for projects before all of the needed revenue has been collected. The feature of having these charges collected by the county as part of the real property tax program will ensure that the RANs will be very highly rated and receive the lowest possible rates.

It should be noted that there are alternatives to the manner in which the STD's charges or fees are imposed. They could be levied per parcel, regardless of size. They could be levied based on square footage of each parcel. They could be levied on front footage of each parcel. Finally, they could be *ad valorem* levies, based on the assessed value of the property.

SEASONAL CHARGES

A seasonal sales tax—charged only during certain months and only in certain coastal counties—is a means of raising revenue from those who enjoy the coasts during those times.

Using Maryland again as an example, the sales tax is 6%. This tax could be raised to 9% from, for example, May to September 15 each year. The point here is that places like Ocean City lure visitors from other states, who get the benefit of the beaches, the dunes, the boardwalks, the amusement parks, and even the well-dredged channels, so they can launch the boats they tow across country. It is only fair that these folks pay some share of the cost of protecting or rebuilding these facilities. Adding another 3% to the sales tax for the big tourism months will enable these people to help pay.

Maryland's total sales tax collections are approaching $4 billion annually. Let us say that 50% of these taxes are collected between May and October. That's $2 billion. Now, increasing the tax from 6% to 9% is a 50% increase. This means that the May–September sales tax collections should go from $2 billion to $3 billion—that's $1 billion a year to help Maryland pay for all of these coastal resiliency projects. Assuming $1 billion is much more than is needed, the tax could be further narrowed by having it charged only in specific counties that benefit from their location on or near the coast.

Will the imposition of a seasonal sales tax create howls from the tax-paying public? Yes, BUT! And the "but" is because sales taxes can be deductible from state income taxes. So, the extra sales tax that residents have to pay can be offset by larger state income tax deductions. That won't stop the howling completely, but it will certainly dampen it.

GEOGRAPHICALLY TARGETED CHARGES

These are charges that a state could impose to help out-of-staters pay their share of coastal resilience costs. For example, there are about 30 million cars that cross the Chesapeake Bay Bridge in Maryland each year. Let us say that only 10% of these cars, or 3 million, are from out-of-state. The state could impose an additional $3 toll on out-of-state cars. They could easily do this especially with E-Zpass. This would raise $9 million "a year" to help pay for coastal resiliency projects.

The same could be done in other states on other bridges and causeways leading to the coast, or on ferry boats. There is voluminous precedent for these targeted charges. In many communities, the use of public parks, or parking at public parks, requires a permit, which is free to local residents but must be paid for by nonresidents.

Over time, states will be required to shoulder more of the burden of paying for coastal resilience. The above strategies are food for thought as to how they might do so.

Section IV

Resources and Disclosure

20 Disclosure

Although municipal bonds are exempt from registration with the Securities and Exchange Commission (SEC), they are not exempt from the anti-fraud provisions of the Securities Act of 1933, including the rules setting out an obligation on underwriters to perform diligence under Rule 10b-5. Disclosure is what prevents fraud. So, disclosure is legally essential to proper bond issuance. In addition, if there isn't good disclosure, the bonds may not be purchased, as investors may feel that their questions are not answered. Additionally, poor disclosure may lead to lawsuits if events turn sour, as investors seek to get their money back, together with damages. So, what would be the point of little, poor, or no disclosure?

Many years ago, I ran New York State's bank for economic development. To finance our loans, we issued tax-exempt municipal bonds. The first time I issued a bond, I went to talk to the State Budget Director to discuss the type of information I should include in our disclosure document. My agency's bonds were guaranteed by a specific provision of the state constitution. So, the state, itself had to prepare its own disclosure document, which is how the Budget Director was involved. However, even so, my agency had to disclose too. During this process, the Budget Director gave me advice that I have never forgotten. He said: "Whenever you are in doubt about disclosing a matter, err on the side of disclosure." I pass this great advice along to you.

Now, let me leaven this statement with some words about materiality.

Let's begin by recalling for whose benefit disclosure documents are prepared. First, they are prepared for the rating agencies that assign credit ratings to your bonds. Good credit ratings are essential to getting the lowest possible interest rates, which translate into the lowest possible annual payments on your bonds.

Disclosure documents are also prepared for municipal bond insurance underwriters, should you choose to have your bonds insured.

Finally, disclosure documents are written for the investors who buy your bonds. If your grandmother buys municipal bonds, she may not read all of the disclosure documents related to a bond she might buy, but you can bet your life that someone at the firm who sold your grandmother those bonds read all of the disclosure documents, for sure.

As far as disclosure to investors or to the general public goes, the vehicle for getting this information out is the Electronic Municipal Market Access (EMMA) platform, or website, that is operated by the Municipal Securities Rulemaking Board.

All of this disclosure may sound daunting. It is, indeed, a lot of work. However, as you will see in Chapter 21, you will have a lot of help from the municipal advisor and the bond counsel you hire to assist you with your bond transaction. Disclosure is very doable.

So, what do you have to disclose? The answer is: "anything that is material to the viability of your organization."

A very wise bond counsel once described it this way. Let us say you are selling a forest. You would certainly say how many acres of land were involved. You would also estimate the average number of trees per acre. You would indicate the general age of the trees, and, of course, you would say what kind of trees they were and that there were no signs of any diseases or other matters that might affect their growth. That's about all.

You don't have to talk about the other plants growing in the forest. You don't have to talk about the bushes and shrubs or about how many blades of grass there might be. My friend, the bond lawyer, once told me that she had seen a summary of a document. The document itself was 150 pages. The summary was 145 pages. "That is not a summary!"

So, please remember that the rule is to disclose any information that was material to the continuing operation of your organization. However, don't hide the important facts with clutter.

As you will see, there are two basic categories of disclosure: financial and operational. The financial side is pretty well-traveled ground. Water and wastewater systems, all have annual audited financial statements. Usually, the financial information in the last 3–5 audited statements will be enough. Be sure to have any supporting information for any footnotes.

The operational side is another story. Let's take an example about staffing.

You would certainly say how many staff were there and what had been the trend over the last 5 years. You would mention the key staff positions and how long most of those people had served in those positions. Do you have to describe each staff position? No. What about turnover? No, not unless there was something unusual about the turnover, especially if the organization itself were responsible, one way or another, for the turnover.

What about a small organization that lost one employee every other year or so? Nope. Nothing to disclose. What if all of a sudden that same organization lost six people last year: One retirement, one death, two moving out of town, and two women who had babies and decided that they wanted to be stay-at-home moms? What to disclose? Is any of this material? Do any of these circumstances have anything to do with the viability of your organization? No, absolutely not. So, should you disclose these matters? I would say that it wasn't necessary, but then, I would remind you of what my friend, the Budget Director, said: "Err on the side of disclosure." So, if it were me, I would disclose. However, I would just say that we had unusual staff turnover last year, but in each case, it was for personal reasons, unrelated to the operations of the organization.

To facilitate further discussion of disclosure, let me introduce you to a very valuable resource. The resource is the National Federation of Municipal Analysts (NFMA). Please visit their website.

As you will see, these are the people whose job is to review municipal disclosure documents. As noted above, they work for the rating agencies, the municipal bond insurers, but, most importantly, they work for the major financial institutions that buy your bonds.

As you will see, they have produced a very useful document called "Recommended Best Practices in Disclosure for Water and Sewer Transactions." It runs only six pages.

Most of this document describes their view on what the content should be of water and wastewater utility disclosure documents. Your bond counsel may say that not all that is in this outline needs to be disclosed in your particular case and that it is a "wish list" of disclosure. Why? Because some of this information may not be material to you.

Here are some of the topics that the NFMA document covers: a description of your utility, including its economic base, its service area, the population trends of the area, and the area's largest employers.

The document asks about peak and average flows, as well as the conditions of the top 10 customers.

It then covers the organization itself, including its management, its staff, and its plant and physical facilities.

It asks what permits the facilities operate under and what their status is. It asks about the environmental regulations that apply to the facilities and what the record of compliance has been. Finally, it asks what—if any—new regulations are expected and how the utility's management intends to deal with them.

It asks what the rates are and how they compare to neighboring systems. It also asks about rate increases—when they occurred—and if any are planned.

It covers the contents of the organization's annual audited statements. It asks about the terms and conditions of the debt that is already outstanding, as well as the debt that is being issued. It also asks about any new debt that is being planned for the future.

Finally, it asks about the bond issuance plan. Most of this information is contained in the document called the Official Statement (OS), which is the most important document in the transaction, since it describes the entire transaction—from disbursement to repayment—in detail. Your municipal financial advisor and your bond counsel and/or disclosure prepare the OS for your bonds.

In summary, disclosure is painstaking and time-consuming, but it is absolutely essential. It is essential to getting the broadest possible market for your bonds with the best possible credit rating. It is also essential for not running afoul of the SEC's anti-fraud rules.

If you think that your disclosure obligations end when the bonds are issued, you are wrong. Disclosure is not over at bond issuance.

SEC Rule 15c2-12 prohibits an underwriter from buying a bond, with certain exceptions, unless the issuer has committed to provide continuing disclosure. Your bond counsel can give you accurate advice as to what this rule precisely requires your organization to do. However, in general, the continuing disclosure rules do require you to post, on EMMA, your annual audit and "financial and operating data" comparable with that included in the OS. So, what you include in the OS is basically required to be updated and posted annually.

Since the SEC felt that issuers were not taking these obligations seriously, they launched the Municipal Continuing Disclosure Cooperation Initiative, which led to the disclosure, by organizations, of their past failures to provide this continuing disclosure.

What if you discover a couple years after issuance that something that you disclosed was inaccurate? As an example, let us say that your pre-issuance bond documents state that the 10 largest employers have maintained a stable employment base for the last 5 years. Now, all of a sudden, three of these companies have announced major layoffs. Is this an event that you need to disclose? Yes, absolutely. These events have a material bearing on whether or not your ratepayer base will significantly decline or whether the ratepayers will be able to continue to afford making their payments. When do you need to disclose? Generally, bond counsel will say that you do not have a duty to speak at all times, but, if you *do* speak, you have an obligation to be complete and not misleading or omit to state something material. So, you may not need to update that employment data immediately on discovering it, but when you next speak to the markets with a disclosure document or continuing disclosure, you should correct the information.

So, as much of a nuisance as disclosure can sometimes be, it is absolutely essential. Organizations absolutely must disclose any events that may have a material bearing on the viability of their enterprise, or especially, on the continued viability of the income stream on which the payments of the bonds depend. It is never pleasant to have to disclose unpleasant information. However, in a free market such as ours, fairness isn't a luxury, it is a necessity. Think of all the grandmothers who could lose a substantial part of their lives' savings because they bought bonds that had hidden flaws and defects that the issuers failed to disclose.

On the other, brighter, side of the coin, look at the benefit to be gained from a better credit rating and a broader market when the news is good and the disclosure is complete. The annual payments will be lower on the bonds and so will be the rates that must be charged.

So, when it comes time for disclosure, just remember the words of my friend, the former New York State Budget Director: "Err on the side of disclosure."

21 The Bond Team

The two most important members of your finance "team" are your municipal advisor (MA), or underwriter, and your bond counsel. That is what this chapter is about.

Let us say for the purposes of this chapter that you work for a large public water authority ("Authority") that is going to issue a tax-exempt municipal bond to finance a large project.

MUNICIPAL (FINANCIAL) ADVISORS/UNDERWRITERS

What is the difference between an MA and an underwriter? There are two general ways for a public Authority to sell bonds: competitive sale and negotiated sale. In the case of a competitive sale, the Authority—with the advice and assistance of an MA—offers bonds to underwriters. The underwriter who offers the lowest interest rate payment wins the bonds.

In the case of a negotiated sale, the Authority not only hires an MA but also an underwriter. The MA generally works with the Authority to organize its disclosure and structure the bond issue. Then they hire an underwriter and turn the issuance over to the underwriter, whose job is to negotiate the lowest interest rate directly with investors.

In the old days, MAs were called financial advisors (FAs). Most people still call them that on a day-to-day basis. However, in response to the fiscal crisis of 2008, Congress passed the Dodd–Frank Wall Street Reform and Consumer Act (Dodd–Frank), which added more regulation to the financial markets than at any time since the Great Depression. One of the major provisions of Dodd–Frank that affects the water and wastewater industry in this country is the requirement that anyone who advises any state or local government on any aspect of their finances must be registered with the Municipal Securities Rulemaking Board (MSRB). Once registered, they are MAs. (I have been calling them FAs for so many years that it is difficult for me to think of them as MAs, but I'll do it.)

Before undertaking any project involving external finance, you would be well advised to visit the website of the MSRB and familiarize yourself with the provisions that relate not only to your MA but also to your ongoing obligation to disclose critical information regarding your finances and operations. You should note that underwriters, too, must register with the MSRB. Most importantly, you should review what might be called the ethics or fairness provisions. As you will see, there are strict rules involving "pay-to-play," which was the unwholesome practice of underwriters and MAs making political contributions or giving gifts to Authority board members or staff, or the even worse practice of staff or board members soliciting money or gifts!

So, the law requires that MAs be registered and that they provide accurate information related to the sale of securities. As you will see below in the discussion of the role of bond counsel, there seems to be a question of whom, exactly, does the bond counsel represent: the issuer or the market? In view of the legal requirement that MAs provide accurate information (to the market), the same question can be asked from them: Whom do they actually represent? The answer is: the issuer. Municipal advisors have fiduciary obligations to their issuers, same as bond counsel, and in the case of a negotiated sale, so does the underwriter.

According to the MSRB, here are some of the duties of an MA in a bond transaction:

- Develops requests for proposals and qualifications for underwriters and bond or disclosure counsel, as well as credit enhancement facilities and investment products
- Assists in developing the plan of finance and related transaction timetable
- Identifies and analyzes financing alternatives for funding capital improvement plans
- Advises on the method of sale, taking into account market conditions and near-term activity in the municipal market
- Assists in preparation of any rating agency strategies and presentations (Read "helps preparing disclosure documents for rating agencies, bond insurers, and investors.")
- Coordinates internal/external accountants, consultants, and escrow agents
- Assists with the selection of underwriters, underwriter compensation issues, syndicate structure, and bond allocations
- Assists with negotiated sales, including advice regarding retail order periods and institutional marketing, analysis of comparable bonds, and secondary market data
- Assists with competitive bond sales, including preparation of notice of sale and preliminary official statement, bid verification, true interest cost calculations, and reconciliations/verifications of bidding calculations, obtaining CUSIP (Committee on Uniform Securities Identification Procedures) numbers
- Prepares preliminary cash flows/preliminary refunding analysis
- Analyzes whether to use investment products such as SLGS (State and Local Government Series are special purpose securities issued by the US Treasury), open markets, and/or agency securities for purposes of investment of bond proceeds

So, as you can see, the MA, indeed, has a dual role on an Authority's bond team.

The MA first advises Authorities on what information they should and must provide for the credit rating agency and then walks through that process with you. Next, the MA does the same thing with municipal bond insurers—helps you prepare disclosure materials and walks you through the insurance underwriting process. Then the MA does the same thing, helping the Authority develop the information that it will disclose to investors. In this latter regard, you will get the opportunity to meet your new girlfriend, Electronic Municipal Market Access (EMMA). As you already know from Chapter 20—Disclosure, EMMA is the MSRB's acronym for the Electronic Municipal Market Access platform or website. This is where you will put all the information that you need to disclose to the public, both before and after the bond issuance.

Finally, in addition to assisting the Authority in preparing all the materials needed to get a credit rating and/or bond insurance, as well as for public disclosure, the MA will also advise the Authority about the bond market itself and on hiring all other players necessary for the transaction.

So, read what the MSRB has to say about this most important member of your bond team, and then, choose wisely.

BOND COUNSEL

The issuance of debt is a transaction that is awash in lawyers. First, there is your in-house counsel, who handles your routine, day-to-day legal matters. These would be personnel matters, any permits you have, and, in general, dealing with regulators.

Next would be your "outside counsel." The outside counsel deals with nonroutine legal matters. Examples might involve purchasing a piece of property or representing the Authority if one of its vehicles was involved in an accident or in the destruction of some third party's property.

Then there is the bond counsel. Of all the players involved in the issuance of a debt instrument, bond counsel is the most important. If you doubt this, you can ask her, and she will tell you. (If you want to know why I am referring to bond counsels as "she" and "her," please see the Acknowledgments section of this book.)

The role of the bond counsel is very special. She represents you, but in a very independent sort of way.

When you issue debt, the instrument will be accompanied by a bond counsel "opinion." The opinion will make two critical and essential assertions. First, the opinion will say that the debt that you have issued is valid and is enforceable against the Authority in a court of law. Second, assuming you are issuing a traditional tax-exempt bond, the opinion will say that the interest on the debt is exempt from federal income taxation. If appropriate, it will also say that the interest is also exempt from state income tax.

In short, the bond counsel is telling the world that your debt is what it says it is: it is a valid debt and is tax-exempt. So, the bond counsel technically works for you, but she is a very independent voice telling the world that the transaction is legal and proper and that the income is exempt from income taxation.

In past years, market participants argued that the bond counsel's client was not the issuer, but instead, it was the "transaction," thus making bond counsel's loyalty to the market, not the issuer. Now, however, it is accepted that the bond counsel's client is the issuer.

Chapter 20 of this book involves disclosure. Most disclosure involves management and money. So, what type of disclosure do you need to do for your bond counsel?

Here is what your bond counsel will do in representing a public Authority interested in financing a capital project with a tax-exempt municipal bond:

1. Bond counsel will examine the state law under which the Authority operates
2. Assuming that the state statute authorizes the creation of water authorities by local county counsels, the bond counsel will examine the ordinance or local law that your county adopted to create your Authority. She will examine the local law to determine that it was in accord, and complied, with the state enabling legislation
3. Bond counsel will then examine your charter to determine whether the charter is in accord, and complies with, both the local law and the state enabling legislation
4. Bond counsel will do the same with the Authority's by-laws
5. Bond counsel will then look at the minutes of the board meeting(s) at which the Authority's bond issue was approved. She will determine that the procedure that the board followed in authorizing the bond issue was legal and proper. Bond counsel will then look at whether any publication of the proposed bond issue, if required, took place. She will determine whether any referendums are required. She will also determine whether the officials who purport to be on the board were duly elected/appointed and whether they were properly sworn in
6. In certain cases, bond counsel will then examine the project and make a determination that your organization has the legal authority to undertake it
7. Since the interest on the Authority's debt is exempt from taxation, bond counsel will examine the "users" of the project (i.e., any party, public or private, having any specified rights with respect to the project), the rate system for the project and/or water system (i.e., whether all users pay the same set of adopted rates, based on usage), and other similar review required under the Internal Revenue Code
8. Based on all her investigations, the bond counsel will then write an "opinion" that will accompany the Authority's bonds
9. The opinion will state that: (a) the bond is a valid debt of the Authority, (b) it is the proper form, as duly authorized by the board, (c) the officers of the Authority who actually execute the bond documentation were duly empowered to do so by the board, (d) the interest on the bond is exempt from federal income taxation, and (e) if the state has a personal and/or corporate income tax, the interest is also exempt from state income taxation

So, you see that the soul-baring that you need to do with your bond counsel involves the laws under which the Authority operates, as well as the procedures it follows in issuing debt, and the nitty-gritty of how the project is operated, in order to qualify the project under the Internal Revenue Code.

Now, the words above describe what a bond counsel does for you. There are fewer than 1000 words here. If you would like to know more, you can go the website of the National Association of Bond Lawyers. Rumor has it that there is a document on the website describing more fully the duties of a bond counsel. It is called *The Function and Professional Responsibilities of Bond Counsel.* Rumor also has it that this document is 56 pages long. Happy reading.

22 Professional Resources

There are six major national associations in the water/wastewater business. Each of them offers a wide array of professional development opportunities, as well as assistance with legislation and regulatory affairs. In no particular order, they are the Association of Metropolitan Water Agencies (AMWA), the Water Environment Federation (WEF), the American Water Works Association (AWWA), the National Association of Clean Water Agencies (NACWA), the National Rural Water Association (NRWA), and the National Association of Water Companies (NAWC), which represents the privately/investor-owned water and wastewater utilities.

THE ASSOCIATION OF METROPOLITAN WATER AGENCIES

The AMWA is an organization of the largest publicly owned drinking water systems in the United States. The AMWA's membership serves more than 130 million Americans.

It is the nation's only policy-making organization solely for metropolitan drinking water suppliers. The association was formed in 1981 by a group of general managers of metropolitan water systems who wanted to ensure that the issues of large publicly owned water suppliers would be represented in Washington, DC. Member representatives to AMWA are the general managers and chief executive officers of these large water systems.

The association represents the interests of these water systems by working with Congress and federal agencies to ensure that federal laws and regulations protect public health and are cost-effective. In the realm of utility management, the AMWA provides programs, publications, and services to help water suppliers be more effective, efficient, and successful.

The AMWA is governed by a 20-member board of directors, which represents all regions of the country. Committees on utility management, regulation, legislation, sustainability, and security provide the expertise to achieve water suppliers' goals, including sustainable operations, regulations based on sound science, and cost-effective laws that support the safety and security of drinking water. A full-time professional staff is located in Washington, DC.

The AMWA hosts both an Executive Management Conference and Water Policy Conference each year. It conducts webinars that are scheduled regularly and publishes five regular newsletters and reports: *Monday Morning Briefing, Water Utility Executive, Regulatory Report, Congressional Report*, and *the Sustainability and Security Report.* It also produces other timely publications.

The AMWA dues are a sliding scale based on the number of customers served.

THE WATER ENVIRONMENT FEDERATION

The WEF is a not-for-profit association that provides technical education and training for thousands of water quality professionals, who clean water and return it safely to the environment.

Members of WEF have proudly protected public health, served their local communities, and supported clean water worldwide since 1928. As a global leader in water sector, our mission is to connect water professionals, enrich the expertise of water professionals, increase the awareness of the impact and value of water, and provide a platform for water sector innovation.

Clean water is integral to the development and care of our communities, economy, and the environment. The WEF recognizes that it takes more than innovative ideas, knowledge, and experience to address the water challenges that continue to threaten quality of life—such as effectively managing water in times of scarcity and abundance, addressing aging infrastructure, improving water quality, and adopting more holistic approaches to water management that focus on recovering and reusing resources rather than simply managing waste.

The WEF and its diverse membership of scientists, engineers, regulators, academics, utility managers, plant operators, and other water sector professionals use their collective knowledge to help address these challenges and further a shared goal of improving water quality around the world.

The WEF's Annual Technical Exhibition and Conference, WEFTEC®, is the world's largest annual water quality conference and exhibition. Featuring the highest-quality training, networking, and business opportunities in the water sector today, WEFTEC hosts thousands of water quality experts and water companies from around the world each year. As the home for stormwater professionals, WEFTEC also hosts the popular "Stormwater Congress," a special conference within a conference that focuses exclusively on this important segment of water sector.

WEFTEC is the primary opportunity that the WEF provides for invaluable peer-to-peer, expert-to-professional, and business networking. The WEF's other training and educational opportunities, legislative/regulatory activities, and volunteer-based activities also promote networking. It includes more than 2500 members who participate in WEF committee activities such as writing technical manuals and books, developing training materials, and helping to develop conference programs and content.

The WEF also sponsors the Water Leadership Institute program that encourages innovation, entrepreneurship, and professional commitment from future water leaders.

To assist members and the overall water quality community, the WEF regularly tracks, monitors, and actively comments on legislative, regulatory, and compliance issues. It co-organizes the annual "Water Week" event in Washington, DC, which connects hundreds of water professionals with members of Congress and federal regulators.

As a leading source of water quality expertise, the WEF is committed to advancing the water profession by providing unparalleled access to the world's best science, engineering, and technical practices in the water environment field. The WEF continuously evaluates its programs and services to align with its overall goal of ensuring clean water for the protection of public health and a sustainable water environment.

To ensure that its resources represent the state of industry practice in all areas of water quality, the WEF publishes more than 190 technical publications, including peer-reviewed Manuals of Practice.

The WEF publishes five magazines and journals: *Water Environment and Technology, Water Environmental Research, World Water, World Water Reuse and Desalination, and World Water: Stormwater Management.* It also publishes four electronic newsletters: *The Stormwater Report e-Newsletter, National Biosolids Partnership e-Newsletter, WEF Highlights,* and *Water Environment Regulation Watch.*

The WEF has a range of dues categories, but the general professional dues are $116 per year. Members of WEF must also join their local member association or chapter, which vary and are set by individual organization.

THE AMERICAN WATER WORKS ASSOCIATION

Established in 1881, the AWWA is the largest nonprofit, scientific, and educational association dedicated to managing and treating water, the world's most important resource.

The AWWA hosts its Annual Conference and Exposition (ACE) every year, as well as its annual Fly-In every spring in Washington, DC, which focuses on legislation affecting the industry.

The AWWA also conducts the Biological Treatment Symposium, the Membrane Technology Conference, the Potable Reuse Symposium, the Sustainable Water Management Conference, the Utility Management Conference, the Water Infrastructure Conference, and the Water Quality Technology Conference. It hosts the Financial Management Seminar and the Water Loss Seminar.

The AWWA also offers a full suite of topical webinars and eLearning courses every year.

The AWWA offers the following member-only periodicals: *Journal-American Water Works Association, Opflow, AWWA Connections,* and *AWWA Water Insider.*

In addition to the AWWA Standards and AWWA Manuals of Water Supply Practice, the AWWA publishes hundreds of books, handbooks, study guides, and videos. It conducts multiple educational events throughout the year, along with a full suite of books, periodicals, webinars, and other learning opportunities available to its members and the water community as a whole.

In addition to the multiple conferences and seminars that the AWWA hosts each year, the association also has 43 local sections that offer members opportunity to meet and network with professionals who live in their area.

The AWWA's dues are broken up into individual and organizational categories. Dues within those categories vary based on organizational size and benefits offered.

THE NATIONAL ASSOCIATION OF CLEAN WATER AGENCIES

The purpose of the NACWA is to provide a voice for public wastewater treatment utilities and stormwater management agencies across the nation—in Washington, DC. The NACWA advocates on behalf of its members on the legislative, regulatory, and legal fronts to ensure that clean water agencies are able to best protect and improve water quality for their communities.

In addition to its annual meeting, the NACWA conducts specialized meetings on areas such as legislation, finance, and technology.

The NACWA publishes *Clean Water Current, Clean Water Advocate, Year-at-a-Glance, The Power of Water* as well as white papers. It also conducts webinars and seminars on a regular basis.

The NACWA partners with AMWA, WEF, AWWA, NAWC, and the American Public Works Association to run the Water and Wastewater Leadership Center, which provides training and classes for utility leaders. The leadership center provides a valuable leadership development program, peer networking, resource center, and mentoring.

The NACWA provides many networking opportunities through its multiple conferences held throughout the year. Members can also network by participating on the NACWA's issue-based committees. In addition, the NACWA provides an online community/forum where members can post questions and receive input from the NACWA community, which many members have found to be very helpful.

The NACWA has multiple membership categories, and dues vary depending on the size of the utility.

THE NATIONAL RURAL WATER ASSOCIATION

The NRWA is a nonprofit organization dedicated to training, supporting, and promoting the water and wastewater professionals that serve small communities across the United States. The mission of the NRWA is to strengthen the 49 affiliated state associations.

The NRWA holds several national conferences each year. The WaterPro Conference is the annual meeting held each year at the end of summer. The Rural Water Rally in February is the industry's premier grassroots legislative effort. The NRWA offers a Water District Finance and Regulatory Issues Forum in June in Washington, DC.

The association also conducts periodic meetings on technical and other specialized topics.

The NRWA offers monthly webinars and publishes a quarterly magazine, *Rural Water*, with a special issue *WaterPro* for managers, board members, and other community leaders.

The Water University Utility Management Certification program recognizes the achievement of individuals who have demonstrated excellence in the management of water and wastewater utilities.

The WaterPro Online Community (www.waterprocommunity.org) allows members to network with utility personnel, access training on new technology and regulations, discuss utility-specific topics with peers, and stay up to date with the new developments in the water and wastewater industry.

Individual membership is offered in the NRWA for $49 per year.

THE NATIONAL ASSOCIATION OF WATER COMPANIES

The NAWC was founded in 1895 in Pennsylvania by a small group of privately owned water utilities. It is now headquartered in Washington, DC, and its members are located throughout the nation and range in size from large companies owning, operating, or partnering with hundreds of utilities in multiple states to individual utilities serving a few hundred customers. These companies don't just sell water as a commodity; rather, what they actually provide are the services required to help ensure safe and reliable water treatment and delivery.

Through their state and regional chapters and committees, the NAWC works closely with regulators and legislators at every level of government and supports public policies that increase investment in water infrastructure. The NAWC focuses on federal legislation that affects privately owned water and wastewater utilities. It also focuses on similar concerns at the state level with state public utility commissions that regulate the private water industry. Public–private partnerships are also a major focus.

The NAWC has chapters in California, Delaware, Illinois, Indiana, New England, New Jersey, New York, Ohio, Pennsylvania, and the Southeast.

The NAWC has publications on the benefits of private water service, government affairs, state utility regulation, environment and sustainability, water infrastructure, security and safety, financial tools and data, and the water-energy nexus.

The NAWC Water Summit is their annual meeting. It also hosts a Water Finance Conference and partners with the other water associations on several events throughout the year. In addition, NAWC actively participates in the meetings of the National Association of Regulatory Utility Commissioners and its regional associations, as well as of individual state public service commissions.

Dues for corporate members are set by the size of the organization. Associate member dues are $450 per year.

Glossary

AMWA: The American Municipal Water Association
AWWA: The American Water Works Association
CIFA: The Council of Infrastructure Financing Agencies
CWA: The Clean Water Act of 1972
CWSRF: Clean Water State Revolving Fund
DWSRF: Drinking Water State Revolving Fund
EPA: The US Environmental Protection Agency
LoC: Letter of Credit
NACWA: The National Association of Clean Water Agencies
NAWC: The National Association of Water Companies
NRWA: The National Rural Water Association
POTW: Publicly Owned Treatment Works
SDWA: The Safe Drinking Water Act of 1974
SFR: Self-Funded Reserve
SRF: State Revolving Fund
WEF: The Water Environment Federation
w/w: water/wastewater

Appendix A: The 28 National Estuary Programs

Albemarle-Pamlico National Estuary Partnership
Barataria-Terrebonne (Mississippi River) National Estuary Program
Barnegat Bay Partnership
Buzzards Bay National Estuary Program
Casco Bay Estuary Partnership
Charlotte Harbor National Estuary Program
Coastal Bend Bays and Estuaries Program
Partnership for the Delaware Estuary
Delaware Center for the Inland Bays
Galveston Bay Estuary Program
Indian River Lagoon National Estuary Program
Long Island Sound Study
Lower Columbia Estuary Partnership
Maryland Coastal Bays Program
Massachusetts Bays Program
Mobile Bay National Estuary Program
Morro Bay National Estuary Program
Narragansett Bay Estuary Program
New York-New Jersey Harbor & Estuary Program
Peconic Estuary Program
Piscataqua Region Estuaries Partnership
Puget Sound Partnership
San Francisco Estuary Partnership
San Juan Bay Estuary Partnership
Santa Monica Bay Restoration Commission
Sarasota Bay Estuary Program
Tampa Bay Estuary Program
Tillamook Estuaries Partnership

Appendix B: The 32 States with Prevailing Wage Laws

Alaska
Arkansas
California
Connecticut
Delaware
Hawaii
Illinois
Indiana
Kentucky
Maine
Maryland
Massachusetts
Michigan
Minnesota
Missouri
Montana
Nebraska
Nevada
New Jersey
New Mexico
New York
Ohio
Oregon
Pennsylvania
Rhode Island
Tennessee
Texas
Vermont
Washington
West Virginia
Wisconsin
Wyoming

Appendix C: Federal Crosscutters

National Environmental Policy Act
National Historic Preservation Act
Archeological and Historic Preservation Act
Farmland Protection Policy Act
Coastal Zone Management Act
Coastal Barriers Resources Act
The Wild and Scenic Rivers Act
The Endangered Species Act
The Clean Air Act
Title VI of the Civil Rights Act of 1964
Section 504 of the Rehabilitation Act of 1973
The Age Discrimination Act of 1975
Equal Employment Opportunity, Executive Order #11246
Section 129 of the Small Business Administration Reauthorization and Amendment Act of 1988
The Department of Veterans Affairs and Housing and Urban Development, and Independent Agencies Appropriations Act of 1993
Demonstration Cities and Metropolitan Development Act
Uniform Relocation Assistance and Real Property Acquisitions Act

Index

D

E

F

N

O

P

Q

R

Z